T0107453

Bird from Hell

SECOND EDITION

Gerald McIsaac

Order this book online at www.trafford.com
or email orders@trafford.com

Most Trafford titles are also available at major online book retailers.

Printed in the United States of America.

ISBN: 978-1-4269-6642-2 (sc)
ISBN: 978-1-4269-6643-9 (hc)
ISBN: 978-1-4269-6644-6 (e)

Library of Congress Control Number: 2011906137

Trafford rev. 04/20/2011

www.trafford.com

North America & international
toll-free: 1 888 232 4444 (USA & Canada)
phone: 250 383 6864 • fax: 812 355 4082

Contents

Author's Note

For more than thirty years now, I have lived with First Nations people, those who were formerly referred to as Indians, Native Americans, or Aboriginals. The people in this village refer to themselves as Dene. We live in the village of Tsay Keh Dene in the mountains of northern British Columbia. It is one of the more remote villages in the country, with a population of possibly two hundred people and almost as many dogs. The older members of the village are referred to as "elders," and they are held in the highest respect. They are the equivalent of those referred to as "seniors" in conventional society. These elders have told me stories that have really changed my understanding of the wildlife that exists in these mountains. They have provided me with detailed descriptions of several huge species, many of which the scientific community believes to be extinct. Although at first skeptical, I had to face the fact that these people speak from personal experience, so that everything they are telling me is true. Their description of the species that matches the woolly mammoth was quite persuasive and forced me to consider the inconceivable. This is one more example of truth being stranger than fiction. I am convinced these species exist, and the public has the right to be aware of this.

In Dene society, the names of the deceased are not to be mentioned, so I do not refer to them directly. I have changed the names of certain individuals who prefer to remain anonymous. The village of Kwadacha is located about seventy-five kilometers or fifty miles north of Tsay Keh Dene. The nearest town of Mackenzie is 350 kilometers or two hundred miles away, and can only be reached by gravel roads.

Because many of the readers will not understand metric measurements, I have given all measurements in their imperial equivalent, except for meters, which is about the same as yards. Similarly, a kilogram (kilo) is about two pounds, and a metric ton is about the same as an imperial ton.

In Canada, First Nations villages are referred to as Reserves, while in the United States, they are referred to as Reservations.

I realize that this is a very serious subject, and I am also well aware that not everyone appreciates my sense of humor. That has never stopped me from cracking some rather poor jokes, because there is really no point in taking life too seriously. After all, I have yet to meet anyone who has ever come out of this life alive. And that is a fine example of a really poor joke. I just find it is best not to take myself too seriously. Consequently, even though this book deals with a serious matter, you will find some terrible jokes as well.

I have recorded the stories that these wonderful people have told me as faithfully as possible, and from their descriptions, I have drawn certain conclusions.

If there is any mistake, it is mine and mine alone.

Chapter 1: Background

It is best to bear in mind that the elders grew up in an environment and time that is commonly referred to as the Stone Age. They were members of a hunting-gathering society. To say that life was difficult would be an understatement. Almost everything they used first had to be gathered or killed. That changed when the Dene met traders from another civilization who offered sacks of flour, rice, and sugar for furs, but the change was not immediate.

There was a cultural misunderstanding, one that persists to this day.

The flour and rice was thrown on the ground, and the sugar was thrown into the fire. They liked to see the sugar sparkle as it was thrown into the flames. The Dene did not know that rice, flour, and sugar were all foods. Still, they knew even less about how to prepare that food. They were trading for the containers, which were very valuable. In fact, they were far superior to the stomachs of the animals, which were the only other containers the Dene used.

It is worth noting that the only way the Dene could boil meat was to fill the stomach of an animal with water, add hot rocks from a campfire, and throw in the meat when the water boiled. They tell me that mountain goat was considered the best for this, because they could get three meals from one container.

This gives you some indication of the world in which the elders lived. To this day, I cannot imagine how they survived, much less raised families. Of course, they were nomads and followed the herds, and their knowledge of the animals in these mountains has no equal. I pay close attention to their stories for that reason.

As best I can gather, there were numerous groups of people who came across the Bering Strait over many thousands of years and walked down this natural corridor, this space between two mountain ranges called the Rocky Mountain Trench. The people with whom I lived apparently came across in the latest migration, possibly within the last few hundred years.

The Dene immediately found that the best areas like the coastal regions, where there was a great deal of seafood and where life was relatively easy, were already occupied. Across the mountains to the east, the prairies were also occupied. In fact, almost all the country was occupied, with the exception of the Rocky Mountain Trench, and the people who occupied those prime pieces of real estate had no intention of sharing with newcomers. In that respect, the situation in those days was similar to that which we have today. All too often, in many countries of the world, immigrants are just not welcome.

As a result, the Dene people were forced to subsist in the trench.

For those who are not aware of the geography, the Rocky Mountain Trench runs from the Arctic to Mexico, and we have the rugged Rocky Mountains to the east and the less rugged Cassiar

Mountains to the west. It makes for beautiful scenery and a rough life. These mountains are not very forgiving. Any moment of complacency can prove fatal. When I first met them, they were in the process of changing from a hunting-and-gathering society of nomadic existence to a more settled lifestyle. We lived in log cabins along the banks of the Ingenika River, and we had no electricity or running water. Fresh meat was available but it first had to be shot. We packed water and cut cordwood for heating, cooking, and cleaning. Clothes were washed on scrub boards in laundry tubs. It was difficult, but we enjoyed ourselves.

The village was not accessible by road, and all supplies had to be either brought in by barge or flown in, which was very expensive. We had a store that mainly sold dry goods, and the mail was flown in twice a month. Luxuries were few and far between.

Of course, things are different now. We now live in a modern village a little north of the "old village," as we call it. Electricity is supplied by generators, and we have running water, a modern school, and a fine store. There is now a bridge across the Ingenika River, and gravel roads connect the villages to towns.

As a result, we have to deal with clouds of dust and restricted visibility in the summer, and in the winter, there are icy roads to avoid. The roads also tend to be quite narrow, only wide enough for one vehicle, with the occasional wide spot in the road every two or three kilometers.

We tend to call them logging roads, which they are, although mining trucks, lowbed trucks, and pickups use them. Not too many cars travel on these roads, or at least they do not run for very long, because such vehicles cannot stand the punishment of these gravel roads.

The roads are maintained by the Forest Service. At the start of the road, signs stating the name of the road, such as the Finlay Main Line or the Russel Main Line, as well as the radio frequency to use are posted. The radio frequencies are generally given names, such as Krause channel or Clear Lake channel, for example. The staff pretty much has to name these stations, because all vehicles that travel on those roads have to use two-way radios. To do otherwise is to court suicide. Usually, there is a sign beside the road with a posted kilometer number every two kilometers or so. Most people use the word click for kilometer.

For example, a logging truck travelling south to the mill with a load of logs may see a sign that reads, "18 km." The driver knows that km is short for kilometers, so if he is on the Finlay Main Line, he will get on the two-way radio and call eighteen loaded on the Finlay. Any vehicles travelling north or away from the mill is expected to pay attention to this and pull over into a wide spot, out of the way. For those who consider this unfair, consider the fact that each logging truck may carry fifty tons of logs, and if the truck comes to a sudden stop, then the logs it is carrying tend to come forward into the cab of the truck. Such things do happen, and they tend to ruin the whole day for the driver.

At the start of the year, this terminology changed. Across the province, the vehicles that are heading into the forest are now required to use the term "up," while the vehicles that are heading toward town are required to use the term "down." The signs that announced this change in terminology were clearly posted near town. For those of us who did not live near town, this change made for a little confusion.

The side roads off the main line lead to logging areas, which we call "logging blocks." After the timber is harvested, the roads inside that block are deactivated. That just means that a machine, usually a backhoe, goes into the area and tears up the road at regular intervals. At that point the logged out area is replanted.

To get to Kwadacha, there is a road called the Russel Main Line. It follows the Finlay River north to the other village, but it only runs on the west side of the river. On the east side of the river, there is another road called the Finlay Main Line. Traffic on the Russel Main Line uses a different radio frequency than traffic on the Finlay Main Line. In each case, the vehicles calling their clicks have to specify which road they are travelling. The empty vehicles may or may not call their clicks.

Recently, a couple girls nearly died when the vehicle they were driving broke down on the road to town. It was the middle of winter, and the temperature got down to forty degrees below zero. They had no two-way radio and were unable to call for help. They spent a very cold night in the pickup, and fortunately, they were found the next day. They would not have survived another night in that severe cold.

Of course, the reason they were not rescued earlier was because no one was looking for them. The first vehicle that came along stopped and gave assistance. The roads in these mountains are not to be confused with highways.

The girls had lived in these mountains all their lives and had driven that road many times without any major mishaps, and as a result, they developed a certain complacency that nearly proved fatal.

Most people here carry a box of candles and matches in the winter months for just such emergencies. Candles throw a great deal of heat, or at least they do if they are lit.

When people like teachers first move out here, they tell me that the first obstacle they face is the culture shock. Apparently, it is similar to moving halfway across the world, so if the reader has a difficult time imagining life in these mountains, you are not alone. I am hoping the pictures I have provided will prove helpful. These mountains are remote and rugged, and they are the home to several species that are thought to be extinct. These species are rarely seen, in part because there are so few people living in these mountains.

I pay strict attention to the descriptions of the species, and I use these along with my powers of observation. These are the facts, and I focus on these facts. These people have also shared their beliefs with me, and while I respect their beliefs, I do not necessarily share them.

I often checked scientific literature and journals for animals that matches the descriptions I have heard. Then to confirm my suspicions about an animal, I tried to speak to people who had seen the species. While this was not always possible, I have spoken to individuals who have indeed seen these creatures on numerous occasions. I always allowed them to describe the experience in their own words. As a result, I have come to conclusions that are stranger than fiction.

I should add that the elders are aware of my efforts and have given them their approval. They are helping me in any way they can.

For that reason, any youngster who tries to lead me astray with a line of phony garbage could be making a serious mistake. That would be disrespecting the elders, and that is never a good idea.

For quite some time, I have remained silent with my suspicions, because I did not want to be locked up as a crazy person. I have since learned that even though many people do not question my sanity, for they are already convinced that I am crazy, there is still no need to lock me up as long as I can function after a fashion. That is a great comfort, because I hate confined spaces. I believe the scientific name for that is claustrophobia.

I did call the biologist in the nearest city, Prince George, and told him about my suspicions. He was skeptical to say the least, which is just a polite way of saying he did not believe a word I was saying. This did not terribly surprise me. In fact, it was the very answer I expected. After I did more research and became convinced of the existence of these huge species, I decided to go public with this book. They absolutely exist, and now it is simply a matter of proving their existence. This is not as far-fetched as it may sound. These species are more widespread than I first thought they were, and I have enclosed clear directions on the method for finding them. I feel a sense of urgency in this, because at least a couple of these species are man-eaters.

The species that terrifies these people more than any other is the one they refer to as the "Devil Bird." They are commonplace in these mountains, and the latest encounter happened about a two-hour drive south of the village of Tsay Keh Dene, in a place we call Raspberry, where a fellow had purchased a homestead recently. He was outside after dark and was attacked by a flying animal. He survived the encounter and now knows that the legends he has heard are true.

His story is typical. People move into these mountains knowing that only superstitious fools believe in spirits flying around in the darkness. Of course, they are half right. It is not spirits they have to worry about but flying animals. As a result, they go outside after dark and frequently pay the price. This man is lucky to be alive today.

The description the Dene people have given me precisely matches that of a pterodactyl. I have tried to hunt one down and shoot it, but I have yet to succeed.

The other species that scares them is the one they refer to as the "Hairy Elephant," which matches the description of the woolly mammoth. Because it is the largest land-dwelling animal in the world and an ornery beast at that, their fear is understandable. Apparently, it did not go extinct. At the same time, the largest canine in the world, the dire wolf, did not go extinct. I can testify to the existence of that animal, because I have personally seen several of them. They call that particular wolf the "Wilderness Wolf." This brings us to the world's largest bear, the mega bear, also referred to as the short-faced bear, which they refer to as the "Rubber-Faced Bear." Those are the three species of megafauna, a term that means "big animals," which the scientists assume went extinct at the end of the last ice age. We will return to that subject later.

The other huge reptile, which is referred to as the "Lake Monster," is also a man-eater. I can think of only one species that matches that description, and that is the plesiosaur.

Last but certainly not least, we come to Bigfoot, otherwise known as Sasquatch. The Dene people are supremely well aware of the existence of this species, and I have had my own run-in with them. I have since spoken to several people who have met them up close, too. I am convinced that they are nothing other than a separate species of human, and I know exactly how to make contact with them.

As for the reader who is skeptical, which probably includes almost everyone, I can only say that such an attitude is scientific, and I welcome it. I only ask that you keep an open mind. I am determined to prove the existence of these species. I have set various traps for the devil bird and the wilderness wolf, and I have tried to get pictures of Sasquatch, without success. At least now I know that which does not work, but now I am convinced that I know how to find them. It just takes time and money.

I might add that the people who are not laughing are those who have spent a great deal of time in these mountains. Almost without exception, they agree that there is something out there, but they just do not know what it is. They are about to find out.

The mountains are extremely beautiful and just as dangerous. No doubt my photography is terrible, but I suggest the reader take a good look at those faces, the faces of those who have lived in these mountains all their lives, and consider the fact that these are the faces of my sources of information. These people are the experts on these mountains.

The idea that there are species that have yet to be discovered should not surprise anyone, at least not anyone who has been following the latest scientific developments. Just recently, new discoveries include a monkey in Burma, a wallaby in New Guinea, a chameleon in Tanzania, and an American turtle much closer to home, just to name a few. Then, too, several years ago, everyone was shocked to find that there were hydrothermal vents at the bottom of the ocean that supported a great number of life-forms, including bacteria, plants, and animals, not to mention life-forms that seem to be a combination of plant and animal.

In my opinion, there other species yet to be discovered here in North America, such as the ones listed. There are some scientists, such as Luis Alvarez, who are of the opinion that the pterodactyl and the plesiosaur went extinct sixty-five million years ago, when a giant asteroid hit the earth and caused such terrible damage that the dinosaurs and reptiles were largely wiped out. Then there are others who have the precisely opposite opinion. They do not question the fact that the earth was hit by a huge rock, but they maintain that modern birds are the descendants of dinosaurs and that birds are in fact dinosaurs.

I hear that scientific discussion has been known to break down, and there have even been cases of violence. I hope this is just a vicious rumor. We are all entitled to our opinions, and those who have different opinions should be respected. I certainly have no wish to get involved in such arguments. I just want to prove the existence of these species and leave the arguments concerning birds and dinosaurs and mass extinctions to someone else. Because there is strong disagreement here, I can only suggest keeping an open mind and examining all the evidence. There can be no excuse for personal attacks.

As for those who object that such ancient species must have gone extinct millions of years ago, I can only draw your attention to such species as sharks, crocodiles, and duck-billed platypuses. Such species are truly ancient and have changed only very slightly over the many millions of years.

The fact is that it is almost impossible to prove that any species is extinct, for extinct species leave no trace of their nonexistence. We can choose to believe that species like the dodo bird and the passenger pigeon are extinct, but there is no way to prove it. On the other hand, there is a very simple way to prove that a species is not extinct. It is simply a matter of producing a living member of that species.

Chapter 2: Bird from Hell: The Devil Bird

Over the years I have spent a great many hours sitting by a campfire, listening to the elders, generally drinking tea or coffee and eating meat. To this day, I never tire of listening to their stories. In the spring especially, we generally go out and shoot a chicken, which is the word we use for grouse, and roast it by the fire. Or sometimes we shoot beaver or moose and feast on that. Not that I would ever think of shooting an animal out of season; however, my friends do, and that is quite legal. Anyway, that is my story, and I am sticking to it.

As someone who is a member of the community—in fact, I married into the village—I was expected to provide meat for the elders, and I did. It may be best if I do not go into any great detail on that subject; however, the elders were accustomed to eating wild meat, and I made sure they were provided with that meat. The only difference is that now there are not as many elders and those who are still with us are getting rather feeble. That comes with age, and as much as we like to think they will live forever, such is not the case.

I was aware that the elders were terrified of spirits, namely those which came out of caves after the sun went down. They have told me numerous stories of shooting moose or some other game animal close to dark and having to abandon their kill because they were being attacked by the devil bird. To this day, the elders never go outside after dark.

Here, we have another example of a cultural misunderstanding. It is my belief that spirits have no physical form or shape and that they cannot and do not kill animals and eat flesh. Consequently, I could not understand their fear of spirits. My understanding of spirits was not their understanding of spirits.

It is their understanding that spirits can and do take physical shape and form, and as such, they are quite capable of killing animals and eating flesh. The spirit that most scares them is the demon that emerges from hell each night and takes the form of a bird.

It finally occurred to me to ask the one basic, most obvious question, the one question no one has ever asked them. I asked them to describe these spirits.

The answer to that question explained everything.

They said that it has the head of an eagle, a body as big as a man, and a wingspan as wide as two moose hides stretched out, and they also said the flapping of its wings sounded like dry hide. Not only that but the tail was as long as a man was tall, and the tail ended in the shape of an arrowhead. The devil bird may sound like a dog barking, a baby crying, or a woman screaming, or it may make other sounds that these people simply cannot describe.

To say that I was shocked is an understatement. It was at that point that I realized that they were giving me a detailed description of a pterodactyl. Either my Dene friends are mistaken or the scientific community is due for a rude awakening. Could it be that the asteroid that hit the earth millions of years ago did not wipe out the pterodactyl?

I have absolutely no doubt of one thing, and that is that there is something flying around in the darkness that is scaring people, and these people are not easily frightened.

They believe that the caves in these mountains lead directly to hell. They tell me that after the sun goes down, they hear a snapping sound, which is the noise Satan makes when he opens the gates of hell and turns loose his pet, the devil bird. Just before daylight, the devil bird returns to hell. Clearly, this animal is nocturnal. This helps to explain the reason we rarely see them, at least in the daylight.

One of the elders, whom I will refer to as Grandma Ruth, says that when the devil bird comes out of the cave, it climbs to the top of the mountain. The elders all agree that if it catches someone outside after dark, then God help that person. The only chance such an individual has is to find shelter, even if it is merely a stand of timber. This makes sense, because this animal has a wingspan of eight to ten meters (twenty-five to thirty-five feet), so it mainly hunts in open areas.

It hunts in much the same way that an eagle hunts fish. They fly in, sink their claws into that poor soul, lift that person up, and carry them away. This is not a pleasant way to die at all. Just as an eagle is more likely to sink its claws into a rather small fish that it has a better chance of carrying away, a devil bird is more likely to attack a smaller individual like a child or small woman.

I should add that the other names for this animal include Satan bird, thunderbird, and demon bird.

Everyone who has heard the flapping of the wings swears that it sounds like dry hide. All agree that it does not sound at all like the flapping of the wings of birds, which of course have feathers. For those who have spent many years in the mountains, the sound of the flapping of wings of birds is quite distinctive. This leads me to believe that this animal does not have feathers and cannot be a bird.

Because the pterodactyl is a reptile, this can make better sense.

I was at first puzzled by the fact that the devil bird climbed to the top of the mountain just after it emerged from the caves. The mouths of the caves opened onto a vertical face, and I was wondering how an animal that size could climb up that mountain with only two legs. A little research clarified matters. The pterodactyl has three claws emerging from each wing, and the scientists think these claws were used for climbing. They say this because the claws are curled like those of tree climbing creatures of the past and present.
It should be noted that fossilized tracks have shown that this animal sometimes walked on two legs and sometimes walked on all fours. When the pterodactyl walks on all fours, it is known as the devil dog.

The next question that came to mind was the following one: Why would the devil bird climb to the top of the mountain?

It climbs to the top of the mountain so that it can become airborne.

It stands to reason that if the animal was to jump out of the mouth of the cave and open its wings at the same time, it would likely fall to the ground. It is very likely a rather heavy animal and may need a level surface to run along flapping those huge wings in order to get airborne. A high elevation would also give it more vertical space to glide.

The top of the mountain on which the devil bird has its nest is flat. I refer to these mountains as perpendicular mountains, and they are quite distinctive. One face of the mountain is vertical, and the top is horizontal. The nesting ground of the pterodactyl is quite easily identified.

This is not to say that the nesting ground of the pterodactyl is the place where the animal builds its nest and lays its eggs. Eggs need heat in order to hatch, and the caves are far too cold. The only source of heat that comes to mind is the sun. The elders assure me that the devil bird comes out of the cave in the summer and lays the eggs. At that point, I go hunting them.

There are only two ways of flying around in the dark. The first is through the use of excellent eyesight, and the other is by echo location.

Bats use echo location, which is similar to radar. It is possible that the devil bird uses echo location to navigate, and of course, we cannot rule out that possibility. People here assure me that the devil bird is usually silent, aside from the flapping of the wings, so if it uses echo location to navigate, it probably emits sound at a frequency that humans cannot hear. It is far more likely that the devil bird has excellent eyesight, and in fact, the scientists assure us that the pterodactyl had excellent eyesight when they believe it lived.

I have been unable to prove the existence of this monster, and the trouble is that time is not on our side, because this animal is a man-eater and the public is not aware of the existence of this reptile.

Much of its prey includes young women or girls, usually First Nation girls from broken homes located on rather remote reserves. These girls tend to run away from home with no money and hitchhike. They are easy prey for two-legged predators as well as the devil bird.

This animal has been around for a very long time, and it is referred to as a successful species by the scientists. In millions of years, it has evolved very little. Like crocodiles and turtles, it survived the asteroid impact that allegedly wiped out the dinosaurs. The difference is that the public is well aware of predators like crocodiles, but they are not aware of the pterodactyl. That is the reason so many people are killed by this animal. It is time we changed that.

Chapter 3: Devil Bird Attacks

There are various stories about this devil bird. None of these stories are terribly pleasant.

The elders surprised me when they told me a story that indicated that this animal had more brains than I would have expected from a reptile. In the spring when they trap beavers and sometimes shoot the animals, they found that the devil bird came around during the night if they shot a beaver during the previous day. They are convinced that the sound of rifle shots alerted the animal to the presence of food. Clearly, the animal has learned to associate rifle shots with food. When the boys did not shoot their rifles, the animal did not bother them during the night.

Grandma Ruth also told me about the time she was attacked by a devil bird in the village of Grassy Bluff at the base of a perpendicular mountain, a village that was built on the flood plane of the Ingenika River. The cabins there were widely spaced apart. At that time, there were no generators, so there was no electricity; however, there was also no background noise, because generators tended to be very loud. Of course, the devil bird had a nest in that mountain, and it frequently hunted in the village below.

In fact, Barbara, my friend who lived there for many years, mentioned that the sounds that the devil bird made after dark just became background noise. I asked her if she ever wondered just what animal was making those noises. To my great surprise, she said she did not.

Grandma Ruth was well aware of the existence of the devil bird, and she knew better than to go outside after dark. The trouble was that she was visiting and lost track of time. It was a beautiful evening, clear and cool, and she was enjoying the walk home. The sun was just going down, and the last thing on her mind was the devil bird. Then she heard the flapping of the wings, and in a heartbeat, the devil bird went from the last thing on her mind to the only thing on her mind.

In fact, she was not at all surprised, but she was terrified. She immediately began running as fast as she could toward the cabin, knowing that she was in serious trouble, especially after she heard the devil bird fly past her, turn around, and then come back even lower this time. It is quite likely that the animal just flew from the top of the mountain and was still too high to sink its claws into its prey. Then again, this may be a standard hunting technique for the devil bird.

Grandma Ruth, however, was still quite a distance from her cabin. She was in an open area, and the devil bird was approaching her again. She could hear the flapping of the wings and feel the breeze created by those huge wings. Even though she was looking death in the face and was absolutely terrified, she did not panic.

There was one slim chance of survival, and she jumped at it. In fact, she literally jumped at it. The only possible shelter was a doghouse, and without breaking stride, running full speed ahead, she hit the ground and slid headfirst into that doghouse.

Of course, the doghouse was occupied, and the occupant was not at all anxious to share its dwelling with anyone else. For that matter, as far as Grandma Ruth was concerned, the dog was free to vacate the premises. A wrestling match in a very confined space between a very big dog and a very small woman soon followed. Neither was going to budge the slightest bit, and neither one did. The dog was well aware that going outside was certain death.

Her husband eventually heard the screaming and the barking and ran outside with a rifle. Grandma Ruth was rescued, but her husband certainly got an earful. She demanded to know why he was not there when she needed him.

Lynn was another girl from the village who had another close call with the pterodactyl during one night. She was driving from Tsay Keh Dene to Kwadacha after dark when she got a flat tire. Of course, she stopped and started to change the tire, and fortunately, she had dogs with her. The dogs gave warning. They were terrified. They were trying to crawl under the vehicle or climb inside. Lynn allowed them to jump into the cab, and she jumped in with the dogs. As she listened, she could hear the flapping of the wings, and she knew she was being attacked by a devil bird.

Immediately, she recalled the stories she had heard as a child sitting on her grandma's knee. It was terrifying to think that the evil spirit that so frightened her grandparents was now attacking her. Worse, the only thing between that monster and Lynn was a thin pane of glass.

A sudden attack by a predator is one thing, and frequently, it is terrible. On the other hand, an attack that goes on and on can be so much worse. Even though Lynn had her initial impulse of panic under control, her imagination ran wild. What if the devil bird could break through the windshield? What did it really look like? How big was it? Did it have teeth? Had it flown away, or was it sitting on a tree branch, waiting for her to step outside the vehicle?

These and a great many other questions raced through her mind as she sat there alone in the darkness for hours, fighting the panic that was trying to destroy her. The temptation to run—the fight-or-flight reflex is so primitive yet so powerful—had to be overcome. To fight the beast was not an option, because she could not even see it, and to run was certain death.

This girl has my utmost respect, because she controlled her fear and her urge to panic, and she did so for many hours. Whenever the fear became too great, she merely hugged the dogs. In some way, this helped calm her down. The dogs may not have provided any safety, but even the illusion of safety was enough to get her through the night.

Of course, the devil bird disappeared at daybreak, and the dogs calmed down. Lynn then very calmly changed the tire and proceeded on her way.

One of the most remarkable stories, which is by now almost legendary, happened many years ago in Kwadacha. The body of a stallion was found in the morning, and it had been killed most savagely. Clearly, it had fought very hard for its life and had died a horrible death.

The carcass was found beside a tree, and the branches of the tree were broken. Not only that but the testicles and part of the intestines were located at the top of the tree.

This horse was not killed by a bear, a wolf, or a cougar. Bears may drag the carcass away and bury it with leaves and dirt, and wolves and cougars just tend to eat their fill. None of these animals carries part of their dinner to the top of nearby trees, and they certainly have no reason to break branches on those trees. The people of Kwadacha were puzzled and quizzed the elders. The elders were not puzzled at all. They just said that the horse had been killed by a devil bird and that the flapping of the wings had likely broken the branches of the tree. They had seen this before, except the prey animals had been moose and elk.

I suspect the animal was in a feeding frenzy, or as the biologists say, something triggered its predatory instinct, which is not terribly uncommon.

There was a similar incident many years ago when two boys were attacked by the devil bird. One of them was the uncle of my wife, and his buddy was with him back then.

The details are a bit vague, but the boys were attacked by the devil bird after sundown. They responded by shooting at the sound that the animal was making. They must have scored a lucky hit, because they could hear the whimpering sounds that the animal made as it started to die.

The remarkable point to this story is that the boys did not go back in the morning to grab any meat. No doubt they regarded this animal as truly filthy, not fit for human consumption. Bear in mind that turtles are also retiles, and some people love turtle meat. I suspect that if it was a turtle, the boys would have eaten it. No doubt it is simply a matter of taste prejudice.

In those days, when people were living on the edge of starvation, such an action was extremely rare, and it gives some indication of the hatred people have for that animal. But people tend to hate that which they most fear anyway.

Several years ago, some boys from town came up to the village to do some work on the store. It so happened that they were amateur climbers, and they liked to climb mountains. I cannot imagine such foolishness, but to each his own. In their spare time, they sometimes drove to the top of Buffalo Head Mountain, which was quite close to the village. At that time, the logging roads had yet to be deactivated. They chose this mountain because it was quite accessible and had a vertical face that they could rappel down. It also had a great many caves that opened onto that vertical face. In short, they chose a perpendicular mountain.

As they rappelled down the mountain, they could not help but notice the bones that were visible at the mouth of the caves. They thought this was rather strange, and once they were back at the village, they asked the people if they had placed the bones in those caves. Of course, they had not, but the people here thought the question was rather strange. Why would they place bones in a cave that only a climber could reach?

In fact, the question only made sense if the bones the climbers saw in the caves were human remains. I rather doubt the climbers would have asked that question if the bones they had seen in the caves had been the remains of animals.

The Egyptian pharaohs of ancient times built pyramids for the burial sites of their deceased, and when these tombs were broken into and robbed, their descendants removed the remains and placed them in caves, ones that were remote and inaccessible. This is commendable and shows a proper respect for their ancestors.

People here also respect their ancestors, and human remains are buried with proper respect. They have no reason to place the remains in caves. I now suspect the bones that the climbers saw in those caves were placed there by the devil bird.

Eyewitness sightings of this animal are rare, but they do happen. They are occasionally seen in the moonlight. I have gone to considerable lengths to confirm that these sightings are genuine and not hallucinations caused by alcohol or drugs. One of the best ways to do this is done by checking with their friends. If the buddies believe this is true, then it likely is just that.

Ben saw a devil bird in the moonlight several years ago. At the time, he was a teenager, and he was just across the road from his house, visiting his cousins. The power went out in the village, and he headed home. The devil bird tended to avoid the light, too. With the power out, the animal moved in. Fortunately, the lad had only a short walk to reach the safety of the house.

Once the lad arrived home, he looked out the kitchen window and saw a huge bird flying around the house he had just left. As he watched, the animal landed on the top branches of a tall spruce tree. It was only much later that he realized just how close he had come to being killed by a devil bird.

This was not the only encounter Ben had with a devil bird. More recently, he was walking in the village after dark, and he felt a breeze on his face even though the wind was not blowing. He suspected it was a devil bird, and he was convinced when he got a glimpse of a wing. By that time, he was aware of the danger of the devil bird and immediately found shelter.

One lad approached me with a scientific magazine in hand and pointed out a drawing of a pterodactyl. He said that the drawing was not entirely accurate, that the beak of the animal was short and stubby. He said he knew this from seeing one of them, sitting on a tree in the moonlight.

There are some people from Kwadacha who had a favorite camping spot on the Finlay Main Line. They no longer camp there, because it is right across the river from Buffalo Head Mountain, a perpendicular mountain. They were all too frequently harassed by the devil bird in the darkness.

The boys also tell me that in the old village at Ingenika Point, a devil bird was chasing them around. Because they could not see it, it was terrifying. They responded by shooting at the sounds it was making in the darkness. In that case, they were not able to land a lucky shot. Partly because they rarely see it, many people think that the devil bird is invisible.

Several years ago, two of the boys from the village were trapping just south of here at Tobin Lake. These two were top-notch woodsmen, and they have lived in these mountains all their lives. The carcasses from the animals they trapped were left in an open area, and that drew some unwelcome attention. Each night after dark, the devil bird would come around. They could not see it; however, they could hear the animal, and the sounds it was making seemed to be coming from all directions around the lake. I suspect it was not one devil bird but a number of them at that time. The harassment got to the point that one of the lads walked out of the area … in the daylight, of course.

Occasionally, the devil bird does land on the ground and walk around. The tracks are quite distinctive, and several people have described them as long and skinny. These tracks are most noticeable in freshly fallen snow. In some cases, the tracks are also described as widely spaced. The devil bird could have left these when it was running along, flapping its wings, and trying to get airborne. Numerous people have seen these tracks, so they must land on a regular basis. I have requested that the next time people see such tracks, they take some pictures.

My friend Vivian recently was walking in the village after dark one night last fall when she saw an animal about the size of a dog. She reasonably assumed that it was a dog until she got close to it. At that point, she realized her mistake.

It was absolutely not a dog, but it was a slimy animal. Just how she could see it as a slimy animal in the darkness is not clear, but I am recording her statement as accurately as possible.

Vivian immediately ran into the nearest house to avoid the devil dog.

One of the lads in Kwadacha once saw a devil bird walking on the bridge leading across the Finlay River. He said that it looked like a man walking in the moonlight.

My friend Brian is from the West Coast, where the symbol of the devil bird is frequently carved into totem poles. He mentioned that the carvings generally included a triangular shape that represented the tip of the tail. Apparently, most tourists confuse this carving with the carvings that represent eagles, or they assume it represents a mythical animal.

There have been reports of people being attacked by the devil bird in the middle of winter, and naturally, I was skeptical at first, because this animal is a reptile and I thought it would hibernate during the winter. Clearly, it does not, because there have been numerous reliable reports of the animal out in weather that reaches twenty degrees below zero Celsius or below zero Fahrenheit. The animal can possibly manage this by a process known as gigantothermia, which is the same process that allows leatherback turtles to survive in the cold environment of the North Atlantic. The reptile loses little heat, partly because a large size leads to a low surface-area-to-volume ratio so that the heat exchange rate remains low. Consequently, the core body temperature changes very little, and a spherical body and a layer of fat also helps.

The pterodactyl evolved between 220 and 240 million years ago on the supercontinent of Pangaea during a time when all the continents were connected. It is entirely possible that as the supercontinent drifted apart over a period of millions of years, this animal survived on separate continents, separated by oceans. It is quite likely that over the period of countless years, some evolution has taken place on each continent, but they may still be recognized as the pterodactyl.

Chapter 4: Hunting the Devil Bird

I was determined to find this devil bird, and because it was a predator and a scavenger, I decided to set a little trap for it. I bought some wire and a ladder, a wilderness camera, and a large piece of meat. I drove out to some private places, which were not hard to do around here, because any place outside the village is wilderness, and used the ladder to run the wire between two trees, and then I hung the meat on the wire. The idea was to hang the meat out of the reach of predators like wolves and bears but within the reach of the devil bird. I also fired off some shots from my rifle, because that seemed to get the devil bird's attention. I was hoping the animal would fly into the wire and break a wing or something, but I had no such luck.

Because that plan was not working, I eventually decided on a more direct approach. The animal would come out of the cave after dark and climb to the top of the mountain, so I figured the place to set the trap was at the top of the mountain where the animal would go over the edge of the cliff.

From the start, things did not work out quite the way I had planned. I chose a perpendicular mountain at Delkuz Lake, and immediately after I left the pickup, I stepped into a hole that was covered with moss. There was a loud snapping sound and a sharp pain in my heel bone. There are seven bones in the ankle, and one of them certainly hurt. On the other hand, I could still walk on it, so I limped up the mountain. I suppose I really should have been wearing hiking boots and not running shoes.

As a result of this incident, the hike up the mountain was a bit more challenging than I had expected. On the other hand, the top of the mountain was pretty much what I had expected, complete with vegetation like willows, grass, and small trees. I was all set to rejoice when I noticed that there was one place that was clear of all vegetation and even devoid of topsoil. It was as if someone had cleared that part of the mountaintop off with a broom. That was the area close to the drop-off, the place where the devil bird climbed over the edge, the place I had planned to set my trap.

The mountaintop was absolutely bare for ten feet or three meters. That spelled the end of my plan to set a trap on the mountaintop for the devil bird. Clearly, they have that base covered. They leave no trails. They merely rip up any vegetation that may get in their way and let the wind blow away the soil. If there is any trail left on bedrock, I cannot find it.

Then again, that bare spot of ground could be a result of natural processes. I just cannot think of any act of nature that would leave that particular area of the mountaintop bare. Either way, whether done by natural processes or not, the result favors the devil bird. I could find absolutely no trace of the animal. My idea of setting a trap was just not practical.

As I stood on that mountaintop, I felt angry. That made no sense, but anger is never sensible. Those animals have been blessed with two hundred million years of evolution, which have given them certain instincts that have served them well. In fact, these instincts have served them so well that most people are not aware of their existence.

For a predator of this sort, which is strictly a stealth hunter, that is the ideal situation.

I put my anger aside, and I faced the fact that my idea that the animal was at its weakest when it first came out of the cave and climbed to the top of the mountain was just not correct.

In scientific terms, we have a name for such ideas. We call them theories, and such theories are meant to be tested. This theory was tested and found faulty. I have been wrong before. I was determined to get over it and get on with my science project.

The only people who never make mistakes are those who never do anything. They merely sit back and watch others try their best, and when a mistake is made, these individuals just shake their heads sadly and say, "I knew that was going to happen." There is no shortage of such people. At least, I have met my fair share of them.

From a scientific viewpoint, this failure just means that a different approach is required. This means going back and once again examining our information. It helps to ask the correct question: At what point is this animal at its weakest? Wonderful things happen when we ask the correct questions. Sometimes we even get the correct answers.

It occurred to me that my mistake was in thinking of this animal as a bird and not as a reptile. There is a huge difference between birds and reptiles. True, both lay eggs, but though birds are warm-blooded animals, reptiles are cold-blooded animals. Both warm-blooded and cold-blooded animals absorb heat from their surroundings, but while warm-blooded animals can increase their internal temperature, perhaps by flapping their wings in the case of birds, reptiles do not have that option. It does not matter how much they jump around. Their body heat will not change in the slightest, unless they move to a different location. They merely absorb heat from their surroundings. This just means that their body temperature will match the same temperature as the room they are occupying, assuming they are in a room. If they are outside in the sunlight, they will similarly soak up heat from the sun.

There are advantages to this. Reptiles require less nourishment than mammals or birds, because they can merely absorb heat. In fact, they can go for weeks without eating, and sometimes they do. Of course, there are disadvantages, too. If they cannot find a readily available heat source, they merely have to stay cold, and their bodies will slow down as a result.

For those of us who think this may be a mistake on the part of nature, think again. These animals have been around for many millions of years, and they are "successful species." Over those many millions of years, they have changed very little.

At the same time I was setting traps for the devil bird, I was also setting a little surprise for the giants and for the wilderness wolves as well. I had my hands full, and in fact, I still do.

Just where were these eggs going to receive the heat they need? Certainly not from their mothers, because the old girls did not have that heat to give. So where else could they get the heat? The sun comes to mind. The sun is an excellent source of heat.

With that happy thought in mind, I asked about devil bird eggs. I figured that an animal that size would lay rather large eggs and that a large egg would show up every now and again, since nature was hardly perfect and not all eggs hatched. To my surprise, I was right.

One of my old buddies mentioned that he had found an egg many years ago. He said that it was brown, that it was about the size of an ostrich egg, and that it had thick skin. He kept it for years as a curiosity and a conversation piece but lost it some time ago. Of course, he had no way of knowing that the egg was worth a fortune.

This was exactly what I wanted to hear, especially when he mentioned the part about the thick skin. The eggs of reptiles have a thick shell. Bird eggs have thin, fragile shells that work for birds, but this trait does not work for reptiles. The shells of reptile eggs have to be thick and not at all fragile.

No doubt we are all well aware of missed opportunities. There is no point in crying about these little things. Hindsight is perfect. The egg is lost, and that is all there is to it.

I knew I was on the right track, so the next question was quite obvious: Do they build a nest in which they lay their eggs?

I approached Grandma Ruth with that question. Clearly, she was surprised, but her answer was clear. She said, "Yes, like a duck nest."

This made my day. For years, I had been trying to figure out just what it was they were telling me. By that I mean just what kind of animals they were talking about and how to prove that it existed. My first thought was that such a task should be very simple and easy. After all, these animals were huge. How difficult could it be to find the world's largest flying animal?

I was half right. Finding these animals is simple, but it is certainly not easy, especially if one is as determined as I am to survive the ordeal. I took some comfort in the fact that I was at least getting ahead of my friends, asking the right questions instead of trailing behind and playing catch-up.

Bear in mind that the nest has got to be built in an open area, one facing south. Only such an area will give the eggs maximum exposure to the sun and receive the heat they need to hatch their young. Not only that but it has to be hidden from predators, too. In short, the nest has to be hidden in plain sight.

The fact remains that the pterodactyl somehow manages to build a nest, lay their eggs, and raise their young without being noticed. At least I am assuming they raise their young. Within thirty miles or fifty kilometers of the village, there are no less than three perpendicular mountains, each home to the devil bird, so they clearly manage somehow.

It is also true that we have yet to find any of these nests, so they must be well hidden from us. The nest must be built someplace we do not go, someplace we do not look. If they were built on level ground, we would have found them long before now.

The devil bird cannot lay its eggs on top of the highest mountains either, because such places are too cold. It follows that they must build their nests and lay their eggs on hills that are fairly small, ones that are not usually referred to as mountains. Around here, any pile of ground less than a hundred feet or thirty meters tall is generally referred to as a hill.

The girls go to mountainsides in the summer to pick berries, but they do not travel to the tops of mountains or the tops of hills, because the berries do not grow there.

It is widely believed that the Dene have travelled all over this country, which is not the case. In fact, we all do the same thing. We stick to trails, and whether that trail is wide enough for one person or a four-lane highway, both are still trails. If we are hungry, we may go to a supermarket or a restaurant. If the Dene people are hungry, they go to a mineral lick or shoot a chicken. We may go to a motel to spend the night, while the Dene may go to a camping spot.

When people think of these mountains, they tend to think of the magnificent scenery and not the smaller hills in the Rocky Mountain Trench. Such hills do not sell postcards; however, they do provide the ideal habitat for the nest of the devil bird.

I suspect the nest is rather small, and the eggs are brown in color, which helps them to blend into the background. I rather doubt the female spends much time sitting on the eggs, because that would defeat the purpose of soaking up heat from the sun. Of course, it is possible that the female abandons the eggs to their fate, but I have it on good authority that she does not. More on that subject later.

It stands to reason that this is a time of great danger for the female. She is outside the protection of the cave and guarding the eggs, very likely hidden nearby, prepared to defend her eggs or fly away and abandon the nest, depending on the size of the predator.

It is doubtful that a person in a low-flying plane would see them, not even if he or she was looking for them. There are thousands of such hills in this area, the Rocky Mountain Trench, and the chances of seeing a nest that blends into the background is not very high.

Then again, it has never been tried, so that may be the way to go. It is worth a shot.

Chapter 5: Looking for Devil Bird Eggs

It was at this point that I realized that my goal of proving the existence of the pterodactyl just became much simpler. Now all I had to do was produce an egg from a pterodactyl nest. As long as I kept the egg warm, it would eventually hatch.

I could well be the first kid on the block to have a pet pterodactyl. This truly falls under the heading of an exotic pet.

All joking aside, there is now something called a DNA test, which can prove to be very useful. Ideally, an egg from a pterodactyl nest could be sent to a university for a DNA test. The results of that test would not prove that it was an egg from a pterodactyl, because no one has pterodactyl DNA now, so there is no way to match it to its prehistoric counterpart. On the other hand, it would not match the DNA of any other known animal, which would prove that there is another species, as yet unidentified, in these mountains.

My questions concerning the pterodactyl nest and eggs received more of a response than I expected. Charity, one of the girls who lived in the village, told me an incredible story.

As a teenager filled with lots of energy, she decided to run up a hill beside the lake we refer to as "Blue Lake," which is the place the whole village went to in the summer for a little swimming and recreation. It was a great spot for the kids to play. Of course, no one ever went to the top of the hill, because there was nothing up there. Everyone knew there was nothing on the tops of these little hills, but everyone was mistaken.

By chance, the hill faced south, and the top of the hill was bare at that time. This was precisely the sort of open area that the devil bird sought out as a place to build its nest.

Charity was surprised to find a big nest there that measured as high as her knees and as wide a desktop. There were also about twenty eggs in the nest. As a curious teenager who thought she was immortal, she decided to investigate and approached the nest.

She was preoccupied, admiring the eggs and considering the idea of stealing an egg or two, when a loud growling and the sound of a charging animal got her attention. From out of nowhere, the animal attacked her. When it was almost on top of her, the animal stopped. For one brief moment, she got a glimpse of what she could only describe as the world's ugliest dog. It was walking on all fours and huddled low to the ground with its head down, snarling and growling the whole time. Her glimpse of this animal was brief, but it was burned into her memory from then on. For months afterward, she continued to have nightmares about this monster.

She turned and ran down the hill and kept running. By her estimation, it was several minutes before she realized that she was still alive, and strangely enough, the animal was not chasing her. Even so, by the time she arrived back at the village, she was still shaking.

Her parents noticed the change in her and assumed the worst. They thought she had been the victim of a sexual assault and was now too ashamed or too scared to speak of what had happened.

True, she had been assaulted, and she had been traumatized. The effect might have even been similar to that of a sexual assault. It is also true that she was reluctant to speak of this, but she had different reasons for not talking. She was afraid that the people would think she was crazy, and in fact, she was beginning to think that very thing. Who had ever heard of someone being attacked by the devil dog and living to talk about it?

It took a while for her parents to drag the story out of Charity, but they eventually managed. Of course, her father was furious and was determined to kill the animal that had attacked his daughter, even if it was an evil spirit, a devil bird. The next day, he charged to Blue Lake with a rifle and his daughter in tow, murder on his mind. Charity showed her father precisely where she had been attacked, and sure enough, the devil bird and the nest were gone.

Her father was one of the finest woodsmen in the country, and he was capable of reading the "sign," which was the word we used for the traces that animals would leave behind, whether these traces were tracks, droppings, scratch markings, or anything else. In fact, such people as her father were capable of reading the sign in much the same way that we read newspapers. The hilltop at Blue Lake was an open book to him.

Even Charity was surprised when her father described the precise sequence of events in great detail. He showed her the route she had taken up the hill, the place she had stopped when she had first seen the nest, and the manner in which she had slowly approached the eggs. He also pointed out the place the devil bird had been hiding and watching her every move. Of course, the animal was hoping she would pass by the nest, but when it was clear that Charity was a threat, it was forced to act. Her father showed her exactly where the nest had been located, where she had been standing when the devil bird attacked, exactly the spot the animal had stopped and barked and growled at her, and exactly the route she had taken, running down the hill as fast as her legs would carry her. The only surprise to her dad was the fact that she had not fallen and broken a leg. Charity then explained to her father that falling and breaking a leg was the least of her worries at the time.

The only mystery was just how the devil bird had picked up the nest and carried it away. The woodsmen here are not philosophers and do not question what they do not understand. They just accept it, and because the devil bird had clearly not walked away with the nest, it must have flown away with the nest. A flying animal leaves no tracks.

Setting a trap for an animal is one thing and has its own hazards. Stealing eggs from the nest it builds is something else entirely, especially when it is the most feared predator in North America. This is not a task for the faint of heart. Still, my mind was made up. I was determined to find a nest and steal some eggs.

The first order of business was to determine the sort of weapon I would carry with me. No doubt I would need a close-range weapon and one that would pack a punch. It was very likely that I would be facing this animal at a range of less than twenty yards.

A high-powered, long-range rifle was just not the ideal weapon. For close-range work, it is hard to beat a shotgun. So I bought a 12-gauge pump-action shotgun and some double-ought buckshot. It was the closest thing to a handheld cannon I could find. I also carried a hunting knife and a short-handled axe on my belt, and I brought a dog with me on my search for the devil bird.

For those who scoff, I can only point out that I went hunting the devil bird alone, because no one else would go with me. There is no doubt at all in the minds of the Dene people that this animal exists and that it is extremely dangerous. They just think that hunting it is a form of suicide. That is the reason why I am armed, why I wear a leather jacket as a form of armor, and why I go hunting during the coolest parts of the day, because reptiles become rather sluggish in the cold.

The fact that each nest may contain as many as twenty eggs, each of which may weigh a kilo, was cause for concern. Eggs are quality food, and they are always in great demand in the wild.

No doubt I was not the only one looking for those eggs. Other predators were also going up those hills with the same idea. No doubt bears, grizzlies, wolves, wolverines, cougars, and coyotes were all hungry for devil bird eggs, and I was certainly not about to forget the other species I was seeking, such as the wilderness wolves and the rubber-faced bears. Those are the species that tend to make the other predators look rather tame by comparison; however, no species can match the giants, otherwise known as Bigfoot or Sasquatch, for sheer ferocity.

The point is that I needed to be prepared for more than just an angry mama pterodactyl.

I could handle the smaller predators with a shotgun and buckshot. Grizzlies and bears would likely not be at all impressed with buckshot, although it could irritate them. I always brought extra shotgun slugs for such situations, and though I was wildly optimistic, I hoped and prayed I would find the time to empty the shotgun of buckshot and reload with slugs if it became necessary.

However, I will never take a shot at a giant. I am convinced they are human and should be treated with all the respect that we would give any other human, if not more so. The fact is that anyone who causes harm to a giant, deliberately or not, would very likely not make it out of those mountains alive. Even if he or she did make it out alive, it is very likely the giants would get even, not necessarily with the offending person.

I am sure that many smaller predators are often scared away by the mama pterodactyl, while the larger ones are not at all impressed.

As a result of this, the female has a choice of laying down her life for her eggs or making herself scarce. I suspect most of them run or fly away from the larger predators. This may explain sightings of pterodactyls in the summer. So too, it is just possible that the pterodactyls take leave of their senses on some occasions.

I also bring my imagination with me, which is an unwelcome addition. I find myself asking, "What if there are a great many of them? What if more than one female is guarding the nest? What if they charge at me on two legs with wings stretched out? What if they come at me on all fours and then take to the air and dive-bomb me?" At such times, I just tell myself to shut up and focus on the job at hand.

In fact, I am not a big game hunter, and I much prefer to see the trophy rack on the animal rather than on any wall. I hunt strictly for meat; however, I admit that I was hunting big game for something other than meat then, and I was enjoying it immensely. It was absolutely thrilling. While I am still not a fan of big game hunting, I can now understand it.

I went up those hills very carefully, and at the crest of the hill, I walked along the edge, keeping my back to the drop-off. I was looking for a nest while keeping a close eye on the dog. No doubt the devil bird builds a nest and then hides on top of the hill, perhaps in a clump of brush. I was determined not to take any chances. If the dog became terrified and started staring at a lump on the ground, I was prepared to fill that lump with buckshot. When it comes to the devil bird, my policy is to shoot first and beg forgiveness later.

I checked out that hill top at Blue Lake as well as several others that my friends told me to avoid. Judging by the vegetation, it was clear that these hilltops had been bare in the recent past. The first plants to sprout are grass, bushes, aspen, and willow, while pine and spruce come later. So my guess was that these hilltops probably contained pterodactyl nests at one point.

I came up empty-handed, and then I took a long look around. There were literally thousands of such hills in this trench, and my chances of choosing the correct one that contained the nest of the devil bird at random were slim at best. No one said that finding the devil bird was going to be easy. If it was easy, it would have been found long before now.

Either way, I had other irons in the fire. I was not entirely focused on the devil bird.

The old cabin I used to live in was still standing, and it was vacant then, so I set a little bait in the cabin for the giants. I also drove some small nails in the doorway, but I didn't hammer them in all the way to the head. They were sticking out just slightly so that if a giant walked through the doorway, it would likely leave some hair behind. I knew exactly the kind of bait that would draw a giant, but I will reveal those details later. I was hoping to get enough hair for a DNA sample, but that did not work out, either.

My friend Fred picks up garbage in the village and frequently keeps dogs with him when he goes to the dump, and that is where the wilderness wolves frequently eat garbage and hunt dogs. My friend is an excellent shot and packs a rifle, but that, too, did not work out quite as planned. We will go into more detail on these topics later.

Chapter 6: Devil Bird and the Golden Eagle

Patience and persistence is the key to proving the existence of these species, and it helps to think of techniques that most people would never consider. It is best to bear in mind that just because no one has ever tried something does not mean it cannot be done. It just means it has never been tried.

Is it possible to get a urine sample from a devil bird?

Francis is a highly respected elder in the village, and at the risk of being repetitive, he is one of the finest woodsmen in the country. I stress the fact that there is no finer woodsman than Francis. To this day, he lives on his trap line in a very fine log cabin. He has little use for the new village and its modern conveniences. Francis knows these mountains and the animals that live in them, predators and prey alike.

His log cabin is located about a twenty-minute drive from the village, and I visit him there on a regular basis. I never tire of listening to his stories, but on this occasion, I had a specific purpose in mind. On this particular day, he was tired from checking his traps—and rest assured trapping is hard work—but he took the time to speak to me.

I asked him about the stories I had heard of the devil bird sitting on the top branches of trees. In response, Francis invited me to look around. The cabin was built in a clearing. All trees close to the cabin had been cut down.

Most experienced woodsmen build their cabins in clearings or at least make sure that any trees close to the cabin are cut down. I assumed that they were afraid that a strong wind may blow a tree onto the cabin, and of course, this is a consideration but not the main one. The fact is that such woodsmen are far more concerned with going outside after dark—perhaps to go to the backhouse—and being attacked by the devil bird.

The devil bird likes to sit on the top branches of spruce trees and do that which it does best—wait. These are reptiles, and reptiles definitely know how to wait. They are supremely patient. They seem to sense that it is just a matter of time before a meal comes prancing down the trail. Assuming the devil bird is sitting on the top branches of a spruce tree situated right beside the cabin, it can dive on the first person to come out the door.

Francis was determined to avoid being pounced on by any devil bird. True, to be outside in the darkness is still dangerous; however, the flapping of the wings makes a distinctive sound, and the breeze the wings create can serve as a warning. Francis can testify to this, because he was once attacked by a devil bird when he was a teenager.

At the time, he and his cousin had become separated from his father. The boys had their .22-guages, which were small caliber rifles. It was winter, and of course, the days were short. His father was a bit careless and let the boys get a little bit ahead of him on the trail.

It was after sundown and almost dark when the boys heard the flapping of the wings. From the sound it made, they could tell that it was flying right past them. The trouble was that it turned around and came back lower this time. It was fortunate that Francis had a rifle. In the darkness, he fired at the sound and hoped for the best.

The devil bird flew off, and when his father heard the shots, he came charging in, afraid for the kids. Sure enough, there in the snow was a patch of yellow. It is possible that the devil bird was wounded and that the blood in the snow appeared yellow due to the poor light, or it is possible that the noise from the rifle shots startled the animal and caused it to empty its bladder. Either way, the devil bird was driven off.

This fine woodsman will also tell you that the devil bird favors certain trees. They are marked with yellow stains at the top of the tree as a result of the animal answering the call of nature. It is not clear if these markings are accidental or if the devil bird marks its territory with a splash of urine just as other animals like wolves or bears mark their territory in the same manner. The only difference may be that one species marks the base of the tree while the other marks the top of the tree.

Assuming that is the case, it stands to reason that the devil bird would return to these marker trees on a regular basis to splash some fresh urine, because the scent of the urine would fade after a few days. Is it possible to cut down one of these trees, strip the bark off the top, and have it sent to a laboratory for a possible DNA sample?

Possibly not, but because the direct approach is not working, I am considering other options. We may never know until we try.

This also confirmed another suspicion of mine, namely that the devil bird is surprisingly light. The top branches of these trees are very small and cannot possibly support any great weight. True, the paws of the animal are long and skinny as people here have described them, and they can therefore be spread over several limbs, which helps distribute the weight; however, the animal must not weigh a great deal even then. This begs the question: How much can the animal weigh?

I can only give a very rough estimate of thirteen to fifteen pounds or six to seven kilos, which I find shocking.

The fact is that the most feared predator in North America weighs no more than a rather small dog. My mistake was in thinking of this animal as a huge, powerful, heavy predator. Huge and powerful it is, but heavy it is not.

It may help to compare the devil bird to another very successful flying predator, a raptor referred to as the golden eagle. This animal frequently hunts mountain goats, and the elders tell me it tends to knock mountain goats off of high cliffs so that the long fall to the ground below kills them.

Then again, there are times when a large golden eagle will sink its claws into a mountain goat, drag it off the ledge, and hang on with wings outspread as they fall to the ground far below. The wings act as a parachute and slow the fall until the eagle can flap its wings and carry the prey to its nest. Considering the fact that a mountain goat weighs anywhere from a hundred to three hundred pounds, it is clear that the golden eagle can carry away a prey animal that weighs many times more its own weight.

True, there is a big difference between knocking an animal off a cliff while hanging on to that animal and picking an animal up and carrying it away. That big difference is the size of the wings. The bigger the wings, the better the animal is able to pick up the prey and carry it away. To put it another way, bigger wings give the animal more lift.

The wings of the pterodactyl are about thirty feet or nine meters across, so that they are four times longer than the wings of the golden eagle. This does not necessarily mean the wings are four times bigger, because the wings of the pterodactyl are probably more narrow, at least toward the end of the wings, but it almost certainly means the wings of the pterodactyl are at least two and probably three times bigger than the wings of the golden eagle.

With that in mind and the fact that nature tends not to give animals more than they need to survive, what possible purpose could a huge wingspan serve? The huge wings of the pterodactyl are necessary to pick up and carry away the prey into which the animal sinks its claws.

It is my estimate that the pterodactyl can pick up prey that is a great deal heavier than it is, which puts such prey in the range of around 110 to 120 pounds, or fifty to fifty five kilos. This is very close to the estimate my Dene friends give me, only they word it differently. They say the devil bird can carry away a child or a small woman.

As for those who object and claim that wings of feathers may be far more efficient than wings of hide, there is no proof of this. Besides, the pterodactyl was flying around millions of years before the first birds took flight.

It may help to think of these animals in terms of military aircraft. We can liken the smaller, faster raptors to jet fighters determined to shoot down enemy aircraft. We can liken the pterodactyl to an enemy bomber, a rather large and slow one capable of carrying a great weight of bombs.

Just imagine a World War II bomber being attacked by a modern jet fighter. It takes no stretch of the imagination to see that the bomber would have no chance. The only way the bomber could possibly survive is to avoid the fighters, perhaps by flying in the darkness.

That is precisely the situation we have here in the mountains. The pterodactyl probably has excellent eyesight, and it is likely right at home in the moonlight. It avoids the sunlight whenever possible, for it is no match for raptors. A fast-flying peregrine falcon would make ground beef of the pterodactyl.

This is not to say that a falcon would tear a pterodactyl apart, because there is no need for that. It would likely just rip open a wing, and the animal would be effectively crippled.

Chapter 7: Fire-Breathing Dragons and Little People

Winter is now coming, and I have yet to prove the existence of the pterodactyl. For that matter, all of the species I am looking for have eluded me thus far. It is also true that these species have eluded everyone else, so proving their existence may be simple, but it is certainly not easy.

This past season, I focused mainly on the pterodactyl because I thought I had the best chance of finding that animal and also because it was a very successful man-eater. The main reason for this success is that it has a great many opportunities to kill people. Most people refuse to believe it exists and put themselves in harm's way.

Just when I think nothing can surprise me, I end up surprised.

One of the girls from Kwadacha approached me with a story that I did not at all expect. I gathered that she had not told this to anyone else, because she thought no one would have believed her anyway.

It happened during a wintery February in the village, and there was a hard crust on the snow. It was after dark, and of course, the village was not well lit. My friend saw some movement in the shadows and assumed that it was one of the boys spying on her.

She was a rather large, powerful girl, and she was not one to put up with such nonsense. She believed in facing trouble head on. That being the case, she charged over to the misguided soul who was irritating her.

As soon as she got close, she realized her mistake. It was not one of the boys, but it was her worst nightmare, a devil bird. She realized she was attacking what she feared the most.

She was not the only one who was scared. The devil bird could sense from the way this girl was attacking that she was a very big, powerful predator. Rather than fight, it decided to retreat, and it did in a manner that left the girl astounded. The devil bird released a cloud of smoke, flapped its wings, and flew away.

Of course, the cloud of smoke was likely a defensive move, designed to cover the retreat of the animal. This is not to say that the smoke she saw was really smoke, and in fact, it almost certainly was not.

We can compare it to an octopus that shoots ink as a defensive mechanism, except the ink is not ink. This ink is actually some sort of substance similar to mucus that dilutes in the water and effectively acts as a smoke screen.

No doubt there is a scientific explanation for the smoke my friend saw, and I wish I knew what it was. However, I know that it was not smoke. This animal does not breathe fire. That is scientifically not possible.

With that in mind, I can only offer a best guess that it releases a substance similar to the substance released by the octopus. This substance could dilute in the air and also act as a smoke screen.

My mental image of this most fearsome predator had to be modified. This animal is almost entirely a stealth hunter. It hunts at night when most animals are sleeping. It usually avoids a knock-down, drag-out battle, if only because it would almost certainly lose the fight. After all, it is quite light, and it must have hollow bones and paper-thin wings. Even a minor injury could cripple this animal to the point where it would be unable to fly, at which point it would be easy prey for other predators.

The fact is that the most feared predator in North America is a fragile animal. The pterodactyl is light. The pterodactyl is delicate. The pterodactyl breaks easily, too.

I am not sure which I find more astounding: the fact that it releases a smoke screen or the fact that the pterodactyl is delicate. Either way, it is best to face the facts and deal with them.

It is also true that the best defense against this animal is a good offense. In other words, if you find yourself outside and attacked by this animal, fight by all means. You have nothing to lose, because it is a predator and determined to kill you after all. If you do not fight, it will certainly do just that, while if you do fight, you have a good chance of beating it off. This animal has good reason to fear you. Just remember that it weighs no more than a fair-sized dog.

For some time now, people in the village have been hounding me to explain the existence of "little people," just as I have determined the scientific names of the various other species, all of which are thought to be extinct. The idea of little humans running around with little bows and arrows and hunting little animals is a bit far-fetched. I cannot see how they could possibly survive, especially up north in this cold climate. They would have to use fire, and as any outlaw can tell you, fire tends to give away one's position.

Without a doubt, the people here believe in these little people, and I made the mistake of dismissing it as just a myth. This fact became clear during the course of a conversation with a fellow from a different reserve.

Most reserves have a clan system whereby a person is born into the clan of the mother. Each and every person in that clan is related by blood, so it is a very practical system. Intermarriage within the clan is usually forbidden. This helps limit inbreeding, the curse of nobility.

Most clans are named after animals, such as Eagle, Wolf, Bear, Grouse, or any other animal. On his Reserve, he mentioned that one of the clans was named after "Little People."

It is my opinion that people are predictable, that we all follow certain habits, and that we only change our habits under great pressure. I have been told that people find a certain sense of security in the familiar, and they will even go so far as to abandon the comfortable to return to the familiar. All too often, people who spend a great deal of time in prison come to regard prison as home, and the idea of serving time in prison no longer acts as a deterrent. I cannot prove this, but these are things that I believe to be true. I have certainly seen numerous examples of such behavior.

As far as I know, all of the clans in Dene society are named after animals. I could be mistaken, and if I am I will be the first to admit it. However, all the clans that people have mentioned to me have been named after animals. Of course, it is possible that one group of people have broken with tradition and named a clan after a mythical animal, but I do not believe this is the case. I now believe that little people do exist, but I do not believe they are humans.

This is not to say there is any law against breaking with tradition, just that tradition can become stronger than any law. I can say that everyone in the village in which I live believes in little people.

But if they are not human, what are they?

Just as the devil bird is not a bird but a reptile, I had to consider the little people with an open mind. To be human, they would have to walk around on two legs. Now what small animal walks on two legs and could be perceived as a small human in the darkness?

The only animal that comes to mind is a newly hatched pterodactyl. Because the adult pterodactyls resemble people when they walk on two legs in the darkness, and because the pterodactyl is a reptile and young reptiles look like miniature adults, I can only assume that young pterodactyls also resemble adult pterodactyls. It follows that a newly hatched pterodactyl would resemble a small person walking around in the darkness.

I mentally kicked myself for not having seen it before. I have always hated it when I have overlooked the obvious.

Of course, after they hatch, the youngsters are probably about the size of an ostrich chick.

Animals are born with certain instincts, and I am sure these freshly hatched pterodactyls are no exception. The fact that they have been around for so long proves that point. I suspect their first instinct is to hide, especially during the day.

So too, they have to eat. The trick is to eat without being eaten. This is truly a tall order when one is born into a world of predators.

Perhaps we can compare these animals to crocodiles, because both are reptiles after all. We may then get a better idea of what to expect.

In the case of crocodiles, the mother carries the newly hatched offspring to the water and guards them for several weeks. Most of the youngsters are still killed by other predators within the first few days, but at least the predators have to work for their meal. Similarly, without the presence of big mama, it is doubtful that any would survive.

Does our resident dragon, the pterodactyl, also stick around and guard her young?

The fact is that nature is only concerned with a system that works. As long as enough animals survive to keep the species from dying out, then anything else is a bonus. We may find this terrible, but nature is only concerned with what works for survival. It is possible that the pterodactyl stays around and guards the young, or she may abandon them. I suspect the pterodactyl sticks around for several weeks after the eggs hatch just as the crocodile does.

The reason I say this is because of all the predators the newborn may have to face. No doubt the wolverine, wolf, marten, weasel, and mink partake in the menu. Various raptors, such as golden eagles, owls, hawks and falcons, may well be their most deadly predators. Larger animals like bears and grizzlies may have a more difficult time, because I expect the youngsters are very fast. Of course, it is possible that the young are able to fly as soon as they are hatched, but I doubt it.

Most researchers, including Christopher Bennett and David Unwin, have concluded that the young, which are called flaplings, were dependent on their parents for a very short period of time while the wings grew long enough and they could eventually leave the nest to fend for themselves within days of hatching. Then again, it is possible they used stored yolk products for nourishment during their first few days of life, as in modern reptiles, rather than depend on parents for food.

Avoiding the light, running around on two legs in the moonlight, I suspect this is the basis of the legend of Little People.

As for dining, I suspect they eat the same things newly hatched birds eat, which consists of a great deal of insects, small animals like mice, and possibly even vegetation.

The most remarkable fact about all this information is that the legend of the fire-breathing dragon and the legend of little people could be based on one single animal—the pterodactyl.

Chapter 8: Highway of Tears

The road from Prince George to Prince Rupert is known locally as the Highway of Tears. It is an eight-hundred-kilometer or five-hundred-mile stretch of highway. Amnesty International estimates that since 1969, thirty-two women and girls, most of them Aboriginal, have disappeared along that highway.

In 2004, they issued a report titled *Stolen Sisters; A Human Rights Response to Discrimination and Violence against Indigenous Women in Canada.* As the title suggests, First Nations women in Canada frequently live in poverty and become the victims of various forms of abuse. It is not surprising that many of the young girls run away from home, and without a vehicle and no money for a bus, they end up hitchhiking.

The Dene girls to whom I have spoken most emphatically agree with Amnesty International's report. Many of them speak from bitter experiences that they have been the victims of abuse in various forms as children, which has led some of them to run away from home at a very young age. While these girls have survived, many of their friends have been killed or have become addicted to alcohol or drugs while they have lived in cities. This is a common result of a life of poverty.

I am of the opinion, which is shared by many of my Dene friends, that most of the hitchhikers who disappear have been killed by this animal. It is also my opinion that many of the people who have disappeared have not been reported.

Of course, it is not only hitchhikers who become prey for the pterodactyl. Anyone in an open area in the darkness is potential prey. This includes people who may just decide to go for a walk. It could include people who have broken down on the highway, perhaps with a flat tire or some other mechanical problem. Then again, there are motorists who occasionally pull over to answer the call of nature. In the darkness of these mountains, that can be a very dangerous thing.

There are a great many Reserves located on or near the Highway of Tears. Most of these Reserves tend to be rather remote and widely spaced. Traffic on the highway tends to be rather light. Aside from settlements that the highway passes through, there are no streetlights. This makes for an ideal hunting ground for the devil bird. Most of the missing girls are in their teens or early twenties and rather slight of build. This is the preferred prey size for the devil bird.

It bears repeating that the devil bird sinks its claws into its prey while the animal is still in the air, picks that prey up, and carries it away. For those who say the highway is well travelled, I can only respond that such an action may only take two or three seconds.

The police have currently classified eighteen of these disappearances as unsolved murders or missing person files, collectively called the "Highway of Tears" case. They have expanded their investigation to other parts of the province and even into the neighboring province of Alberta. The mountains extend into western Alberta, so this is not too surprising.

I cannot even begin to imagine the suffering endured by the families of those missing girls. The last thing I want to do is add to that pain. Neither do I want to see any more girls go missing in these mountains. The fact is that as long as people who live in these mountains go into open areas outside after dark, then such people, especially children and women, will continue to disappear.

Of course, the police are not giving out too many details, but I suspect that many of the remains were found with strange mutilations and claw marks. Some of the bodies are found within a day or two, while others have not been found for months or even years. And of course, some bodies have yet to be found.

There have been unconfirmed reports that some of these bodies have been found complete with money and identification. Because rapists and murderers generally help themselves to trophies, especially money, it is unlikely that these victims were killed by men.

It is not too surprising that some bodies are found immediately. Vera was one of the girls in the village, and she was at home inside the house after dark when one of the dogs outside gave out a yelp and howled as it was picked up and carried away. Vera knew it was being carried away because the sound the animal was making became ever fainter until she could no longer hear it. This all happened very quickly, and in the morning, the body of the dog was found just outside the village. The claw marks of the animal were quite distinctive. The devil bird had apparently lost its grip, and the dog had fallen to the ground. The trauma of the claws as well as the fall from a height was fatal, and this could be confused with homicide in the case of humans.

There are reports that many members of the RCMP (Royal Canadian Mounted Police) are involved in this investigation. This police force is considered to be one of the finest, and in fact, the members are well trained. If they are under the impression that many of these girls have simply disappeared into thin air, it is understandable, because that is very close to the truth.

This is not to say that all of the girls who have disappeared from the highways are victims of the devil bird. Without a doubt, there are other predators, and of course, they prefer the weak and helpless. These human vermin tend to rape and murder their victims, and they are the responsibility of the police.

It is also true that animal predators are more likely to attack if they smell blood. There is a reason so many girls of childbearing age are attacked by wild animals. There is a certain time of month when it is absolutely not safe for such girls to wander in the wilderness. It is quite likely that the devil bird has a fine sense of smell, and without a doubt, wolves, bears, and grizzlies have a superb sense of smell.

The police will have their hands full trying to separate the victims who have been killed by the devil bird from those who have been murdered. Both are the victims of trauma, but at the moment, this is not an issue, because the police are either not aware of the existence of the devil bird or simply consider it a myth.

My Dene friends assure me that the devil bird lives throughout these mountains from the Arctic to Arizona and possibly farther south. As mentioned previously, they have a huge wingspan and prefer to hunt in open areas, and a highway is definitely an open area. It is my sincere hope that the awareness of the existence of this huge predator in these mountains will cause people to think twice before they hitchhike or venture outside after dark.

At the very least, if you are outside when darkness descends, then try to find shelter, even if only a stand of timber. There is no safety in an open area. This is not to say that people are completely safe in heavy timber, either. This animal can and does on occasion land and walk around, sometimes on two legs and sometimes on four legs. It is a predator, one that is always dangerous and never completely predictable.

I have no doubt that people are also being killed by this reptile in the Yukon, Nunavut, the Northwest Territories, and anywhere else these mountains are present. That includes Alaska and such states as Washington. Rest assured, the pterodactyl has killed far more people than any serial killer, and they are continuing to kill today.

No doubt our American friends are facing the same problem with hitchhikers disappearing along highways in these mountains, and they have likely come to the same conclusions the Canadian authorities have reached.

My friend Brian assures me that he became the prime suspect in a murder investigation when a friend of his disappeared one night. Because he was apparently the last person to see that girl alive, the suspicion naturally fell on him. Murder investigations tend not to close, so it is very likely that he is still the prime suspect.

Brian is of the opinion that a devil bird was likely responsible for the disappearance of that girl. I rather doubt that the police are terribly impressed by that explanation.

There are many impressive mountains in Alaska, and in fact, the devil bird is no stranger to Alaska. It is almost certainly present across the Bering Strait in Asia, and in fact, I have had people from Asia mention an animal that matches the description of the pterodactyl. The only difference is that this animal is called the dragon in Asia. The Chinese have even named a year after the dragon. Not unreasonably, they call it the "Year of the Dragon."

For those who say that the dragon of China is a myth, I can only say that the Chinese have named twelve years after twelve animals, not eleven animals and one myth. I doubt the Chinese are different from people anywhere else in the world, and because eleven years were named after eleven animals, I doubt they would break with tradition and name year number twelve after a mythical animal.

That so-called mythical dragon of Asia is the same so-called mythical devil bird of North America. Here, however, we tend to call it the thunderbird, the devil bird, the Satan bird, or the demon bird. But it is not a myth. It is a superb predator. It is the terror of the night sky. It is a man-eater. It is a pterodactyl.

Chapter 9: Lake Monster and the Hairy Elephant

My friends Roy and Kathy assure me that there is a second huge predator located in Takla Lake as well as other lakes in the province like Okanagan Lake. The elders around Takla say that the animal goes into underwater caves around the lake.

I suspect it is a reptile. Those animals travel to those underwater caves for a reason.

These reptiles are right at home in the water; however, they still have to breathe, and breathing in an underwater cave has the advantage of safety. There are very few predators in underwater caves. Of course, it is also a safe place to sleep and keep a nest. The stories the people tell me indicate that this is a huge, powerful animal.

August is the time of year that the bull moose are at their finest. They eat and fatten up all summer in preparation for the mating season of the fall. The trouble is that in the fall, when they go into the rut, they quit eating, fight other bulls, and lose all their fat. For that reason, people here hunt bull moose in August before the rut.

The boys at Takla shot a big bull moose one August day and decided to use a motor boat to tow it across the lake.

No doubt the bull moose was very large and very heavy, because large and heavy tend to go together. Likely, the boys thought to drag the moose to the other side and let the girls butcher the animal. They had a motor boat, so why not use it? And who better to do the hard work of butchering than the wives? And rest assured that those girls knew how to butcher a moose.

A fine plan, but it did not work out quite the way they had in mind.

Partly across the lake, something grabbed that huge moose and took it underwater. Worse, it was dragging the boat backward at a rather alarming speed. The boys in the boat did the smart thing and cut the rope. As a result, they did not get wet. They saw the water swirling around but never saw that moose again.

The animal that dragged the moose underwater must have been very big and powerful.

Even though the First Nations people there are aware of the creature, they have apparently not given it a name. Most people just refer to it as the lake monster, which is a name in itself, but there are some people who refer to it as ogopogo.

From the description that has been provided, I suspect the animal is a plesiosaur. Perhaps it is best to think of this animal as a huge turtle with a long neck, one without the shell. It is a very large predator, possibly fifty feet or fifteen meters in length. It, too, is a man-eater, another prehistoric reptile from the age of dinosaurs that supposedly went extinct along with the others.

It is another predator that is superbly adapted to its environment, likely one that has changed little from the appearance it had millions of years ago. That makes it supremely dangerous just as the pterodactyl is extremely dangerous. I suspect that its main prey is fish, but I am sure that it will eat any prey that comes into the water.

There have been other encounters with people, and not just by the First Nations people. Several years ago, there were a couple professional scuba divers doing work for the government in Takla Lake. The word was that those boys came out of that water a lot faster than they had gone into it, and furthermore, they had no intention of returning to that water after the event. They said that there was something in that water, and it was clearly huge and fearsome. I gather that the language they used left no room for any misunderstanding.

This did not stop the younger people from swimming in Takla Lake. Clearly, young people are the same everywhere. They believe that the older we get, the stupider we get.

There is another story about one girl who was wading on a tire tube in the lake and saw something that terrified her. She, too, vacated the lake in record time. Apparently, the creature resembled a huge snake.

In the spirit of being open-minded, I will suggest another theory here that this animal could be a primitive whale called *basilosaurus cetoides*. That is certainly possible, and as whales are mammals, they, too, have to breathe. It is also true that primitive whales had legs, so the footprints of this elusive monster may be seen on occasion. The trouble is that footprints can easily be faked.

I suspect it may be the same animal in Europe that they call the Loch Ness Monster.

One way we could find this animal is by trolling the lakes for its presence, not unlike how we would troll for a very big fish. A wild animal is a hungry animal, and with a little luck and a very big hook, we may be able to catch a very big reptile. Then again, there may be an easier way to prove the existence of that animal.

The plesiosaur is a reptile, and as previously mentioned, reptiles lay eggs. Furthermore, eggs need heat in order to hatch. That heat is almost certainly provided by the sun, and I suspect the plesiosaur acts in a manner similar to the leatherback turtle. The female of that species comes out of the water, digs a hole in the sand, and lays her eggs there. She then covers the eggs with sand and abandons them.

If I am right and this animal is a plesiosaur, then proving its existence may be a simple matter of finding the eggs it lays. Bear in mind that simple is not the same as easy.

Just as the pterodactyl builds its nest and lays its eggs on the top of hills that face south in order to get the maximum exposure to the sun, the plesiosaur lays its eggs in a sandy location facing south as well.

It is just possible that the eggs of the plesiosaur have never been found for the same reason pterodactyl eggs have never been found. No one has ever looked for them. Or it could be that someone has found an egg or two and put it away in a closet and forgotten all about them. It could just be that simple.

The best way to find the plesiosaur is also the best way to find the pterodactyl. Quit looking for the animal and start looking for the eggs that the animals lay.

My mother-in-law, who has since passed away, grew up in this country at a time when this location was extremely remote. She lived her whole life in these mountains, and in those days, there were no schools. She spoke broken English and never even set foot in town until late in life. That event was an absolute culture shock.

I was just as shocked when my wife told me that her mother had recognized a picture of an elephant. The only difference was that the elephant her mother knew about had really long hair.

That was the day I realized that I was missing something really big, and I mean *really big*. I am talking about the woolly mammoth. In fact, the so-called hair on the woolly mammoth is about three feet long or a meter long.

It takes a great deal to scare the elders. I admit that I did not think it was possible.

That was before they told me about the devil bird and the hairy elephant.

The woolly mammoth scares them. The woolly mammoth is not a pleasant animal. I am sure it is a vegetarian, but it appears to have a strong dislike for people.

There are some scientists who think that intelligent animals are capable of such emotion as hatred. If that is the case—and I believe it is—then as an intelligent animal, the woolly mammoth must hate us. I suspect there is a good reason for this hatred, too.

My First Nations friends tell me that the mammoth was alive in the Midwest in the eighteenth century. They also tell me that it was alive in Alberta in the nineteenth century. And it was definitely alive in this province in the twentieth century.

If I am right, the mammoth was widespread across North America at the time of the European Invasion. As settlers pushed westward, they pushed the mammoth ahead of them. This is completely understandable, because those people all planted crops and no farmer wanted a five-ton vegetarian in his or her garden.

Farmers in Africa are familiar with the problem, and I am sure some of them would love nothing better than to shoot every elephant that came onto their farm. Completely understandable, but they are restricted from doing so by current laws.

Many years ago, the farmers of North America were not hampered by those laws. I am sure they shot the mammoth at every opportunity. It is quite likely that the mountains of western North America contain the remnants of a once-great herd of mammoth.

The elders assure me that if the mammoth catches your scent, it is likely to run you down and kill you. Your only defense is to run for the nearest swamp. That animal is very heavy and cannot go into swamps, for it will just break through. In winter, this is not an issue, because those animals will probably venture into caves for warmth.

My wife tells me that her grandfather once found the remains of a mammoth. The animal had used its trunk to push together trees and form a house, and there were a lot of mammoth droppings in the area, too.

Of course, the scavengers have been busy at the remains for quite some time. No doubt the animal had taken sick and had constructed a shelter for itself. After some time, it had died. I am guessing this happened in the early twentieth century.

More recently, in the spring of 1965, two of the boys from Kwadacha were trapping beaver when one of them saw a mammoth and immediately ran back to his buddy. Both ran for their lives afterward. They figured that the weapons they were carrying wouldn't bring down a mammoth, so they didn't attack the creature. This happened north of Kwadacha at a lake the locals call Guitar Lake.

Those two individuals are no longer with us, but they told some lads who in turn told me that story. There are other stories of mammoths terrorizing people, and it is possible that there have been more recent sightings. I am currently trying to track down some people who have reported seeing animals that match its description.

The fact is that the woolly mammoth is not extinct, and as soon as we prove that fact, it will probably be declared an endangered species. As such, it will be protected by international law. The pressure will be off the animal, and it may even sense this lack of pressure and return to its old stomping grounds.

It should not be terribly difficult to prove the existence of this animal, because they are so big in size. They can probably be spotted from satellite photos, if anyone would bother to look.

For the last many years, this creature has been scratching out a living in the mountains. By now, the survivors may be a rather stunted, sad-looking bunch, half starved, somewhat ragged. That will change as soon as they come out of the mountains. They will find themselves in a land transformed, a land of milk and honey, at least from the standpoint of the mammoth.

The animals will gaze at fields of wheat, rye, oats, barley, and corn that stretch as far as their eye can see. And rest assured that the mammoth can see very far. And just as surely, they will feast on this banquet, because the fence that would stop a mammoth has yet to be built.

Chapter 10: Rubber-Faced Bear and the Wilderness Wolf

Archie is an elder in the village of Tsay Keh Dene. As a teenager, he and a few of the boys found sign of a huge bear. It was clearly a monster of a bear.

All bears mark their territory by standing on their hind legs, reaching up as high as they can and scratching a tree. This serves as a warning to other bears to stay away, and it probably keeps fights to a minimum. After all, there is no point in picking a fight with an adversary who is far bigger than you are.

In this case, the boys were amazed to see that the claw marks were so high up on the tree. They were accustomed to seeing big bears, but this one was clearly record size. The tree was a huge pine, yet the bear was so big and strong that by scratching and pushing on the tree, part of the roots had been exposed. Sometime after that, the boys shot a huge rubber-faced bear.

If took fifteen shots from a .30-30 rifle to bring it down. Bear in mind that at close range, this rifle is very powerful.

They called it the rubber-faced bear because it had no hair on its face or the top of its head. The elders think that while it lay in caves for the winter, the bear rubbed its face with its paws, and all the hair came out. While that is an interesting theory, and I certainly respect the opinion of the elders, it does not explain the huge size of the bear.

I have a different opinion. I think it is merely a separate species of bear, one which supposedly went extinct ten thousand years ago.

The mega bear, which is also known as the short-faced bear, matches the description of the rubber-faced bear, which is also known as the beaver-eating bear by the Dene elders.

This bear is so big and powerful that it is capable or tearing apart a beaver's house. Of course, there are many names for this one species of bear, and then there is a proper scientific name. The trouble is that such scientific names tend to be unpronounceable.

There have been other recent sightings, too. In fact, one of the boys spotted it several years ago on his trap line in the middle of winter while he was hunting a beaver.

The beaver did what beaver always do, namely build a dam in the summer months, flood out an area, and build a house in the middle of the pond that they have just created.

They cut down willow, aspen, and birch trees, which they then use as building materials, and of course, they also eat the bark. The aspen and birch trees are called hardwoods. This does not mean the wood is necessarily hard, although the wood of birch is indeed very hard. It just means these trees have leaves, while trees without leaves, such as pines and spruces, are considered softwoods. These trees are easily spotted because they have needles.

The beavers have to work very hard to build the dam and the house as well as secure a good supply of food for the winter. The entrance to the beaver's house is underwater, too.

The beaver also has to store up a good supply of food for the winter. This means a great deal of wood also has to be placed underwater. If the beaver run out of food, the bark on the wood it has stored underwater, then the animal faces starvation, because it cannot break through the ice to get to any more bark.

In the winter months, the pond freezes over, and the beaver stays in the house under the water. If the beaver is caught outside the pond when the water freezes, it faces death by hypothermia or predators. If all goes well, the beaver stays in the house, safe and sound for the winter, and it never comes out onto the ice until it begins to thaw. At least, if it has done its job properly, it remains safe in the house.

Not all jobs are done properly. Even beaver make mistakes.

So too, dams break, or they are sometimes broken. Water may seep away. Any number of things may happen, all of which could make the life of the beaver miserable. Most people can understand that.

In this particular case, the life of this beaver was made very miserable indeed. It happened during the middle of winter, and of course, the pond was frozen. Naturally, the beaver was safe and warm inside his house. The beaver's house is an engineering marvel, and it is very solid. Even the biggest grizzlies have a very hard time tearing apart a beaver's house. Most grizzlies do not even try.

In this particular case, the mega bear was not at all impressed by the strength of the beaver's house. He just proceeded to tear that house apart as if it was made of straw.

No doubt the beaver ducked into the water to escape, but it had to come up for air at some point. The trouble was that there was only one place to come up for air, and that meant going right into the jaws of the mega bear.

It just so happened that this was happening on the trap line of the uncle of my wife. He was tempted to shoot that bear, but his wife advised against it. She pointed out that he had no one there with another big gun, as people here call rifles larger than a .22-guage, and she was concerned that the bear may decide to have them for lunch instead. That is my idea of a smart woman.

This bear may well be the world's largest bear. A full-grown mega bear may stand six feet or 1.8 meters at the shoulder and weigh close to a ton.

It is possible that the widespread killing of beavers as a result of the fur trade reduced this bear's numbers. With less prey, it is possible the population of bears that preyed on those beavers was reduced. Now that far fewer beavers are being killed, perhaps the mega bears are making a comeback.

This brings us to the wilderness wolf, almost certainly the dire wolf, the world's largest canine. The First Nations people are well aware of this animal, because it comes into the village after dark and kills dogs. Several people have spotted that animal. They tell me that it is even taller at the shoulder than a deer. I can testify to that fact, because I have seen several dire wolves, the largest of which was the size of a caribou. Rest assured that this is one huge wolf.

These three species of animals are referred to as megafauna and are considered to have gone extinct at the end of the last ice age ten thousand years ago, along with the saber-toothed cat and the giant ground sloth. The theory is that a drastic climate change caused their extinction.

It takes many millions of years for such species to evolve. These species were alive at the beginning of the first ice age and at the end of the last ice age. The fossilized remains leave no doubt of that fact. It is also a fact that in the last hundred thousand years we have had three ice ages. It logically follows that these species of megafauna survived five previous episodes of drastic climate change, twice in the form of warming and three times in the form of cooling. This would seem to indicate that these species are quite capable of surviving such drastic climate changes.

I should add that the First Nations people to whom I have spoken have no knowledge of the saber-toothed cat or the giant sloth. I suspect that those animals are indeed extinct, but I have no idea just what exactly caused their extinction.

Then again, they may well not be extinct. Just because people here do not know about them does not mean they are extinct. It just could be that they are very much alive in other parts of the world.

My friend Fred recently had an encounter with a pack of wilderness wolves. Fred is semiretired and makes a little extra money working part-time picking up trash in the village and taking it to the dump. The garbage is not incinerated but buried in a landfill, which makes for a rather smelly mess. This also tends to draw scavengers that see the place as a regular banquet. At any time, there are a great many ravens at the dump. The ravens, however, are not the only scavengers at the dump. There is the occasional timber wolf seen there, and from time to time, the boys set traps for them or sit by the dump close to dark with a rifle, hoping to shoot one.

For several years now, the routine has never changed. Fred has two dogs, and they generally run alongside his pickup, which is great exercise for the dogs and excellent company for Fred. He picks up the plastic garbage bags from each house, and when the pickup is full, he runs to the dump. He then throws the trash into the pit, and once a week or so, the garbage is buried.

This all changed the day he was attacked by a pack of wilderness wolves.

He was visiting the dump and throwing out the garbage when it happened. The wolves were only several meters away before he noticed them. It was almost too late, but thinking quickly, he grabbed a bag of trash in each hand and started waving his arms. The idea was to appear larger than he was and hopefully intimidate the wolves. To his great relief, the wolves backed off. They were temporarily distracted but not beaten. They were not about to give up quite that easily.

It was fortunate that Fred was right beside his vehicle so that he could jump into the cab. His dogs were still outside, and the wolves were still determined to have a meal, not just the scraps from the dump. Because they could not have Fred, they decided to eat his dogs.

From the safety of the vehicle, my friend charged the wolves. His dogs stayed very close to the vehicle, and each time the wolves attacked the dogs, Fred attacked the wolves with his vehicle. For several minutes, it was touch and go, the pickup going backward and forward to save the dogs, until finally, when he was able to maneuver out of the dump and get halfway back to the village, the wolves gave up.

As a result of this, my friend got angry. He did not appreciate having his dogs attacked by wolves, and he sure wasn't happy about having wolves attack him. One good turn deserved another, so he went back to the dump, only this time with a rifle, a .22 Magnum. Sure enough, the wolves were still there and still hungry, but he gave a couple of them a meal they definitely did not want—hot lead from his rifle.

Fred is one of the finest marksmen in the north, and he consistently wins the shooting contests we have here in the village. Most of us still compete, hoping to at least come in second. The idea of beating Fred in such a contest is hardly considered by most, but it is fun to see how close we can come. We always have a good time.

Shooting those wolves was almost too easy, and he managed to get two shots off before the pack disappeared into the woods. Even the two he had shot in the chest managed to make it to the safety of the trees.

Fred did the smart thing. He let the wounded wolves run away, knowing that wounded animals tended to run far when they were chased, while if left alone, they generally lay down and died. Better to come back in the morning and pick up the bodies.

It was a fine plan, but as with most plans, things did not work out as intended.

The wolves he had shot were wilderness wolves, not timber wolves. They are completely different animals and very tough, as he found out the following day.

He returned the next morning with his rifle, all set to finish off the two he had shot just in case they had not died. In fact, they were very much alive, and if the bullets they were packing were slowing them down, it was not noticeable at all.

The wolves hid in thick brush, and as Fred approached, they commenced to bark and howl and lead him away. Following them was easy, because they were making such a racket, but at the same time, they were staying just out of sight so that it became impossible to get a clear shot at them.

For an hour, this cat-and-mouse game went on until Fred realized that the wolves were trying to turn the tables on him. It was not so much that he was hunting the wolves as they were hunting him. Their plan was to lure him to a location of their choice where the whole pack could pounce on him, likely a place of thick brush where his rifle would be of little use. Fred very wisely decided to go to the relative safety of the open road and walked back to his vehicle. It is best not to underestimate these wilderness wolves.

The scientific name for these animals is dire wolf, and dire just means danger. In this sense, the scientific name is accurate, because these wolves are extremely dangerous.

These animals occupy an important place in the wilderness. It is quite possible that they help keep the timber wolf population in check.

The events at Yellowstone National Park provide a clear cut example.

There was a time not too long ago when wolves were considered merely vermin, animals to be killed as quickly as possible. "The only good wolf is a dead wolf" was the attitude. The wolf exterminators did a fine job of killing wolves in Yellowstone. In fact, they did such a fine job that the wolves in the park were completely wiped out.

The idea was to provide a paradise for grazing animals, such as bison and elk, and they succeeded. However, the elk wasted no time in destroying paradise.

The change was rather gradual, so the people running the park were not aware of just what was happening at first. It was only after a period of years that it was clear to even the most stunned park wardens that the park was being turned into a desert.

The biggest offenders were the elk. They were eating all the young vegetation, even the young aspen trees. Of course, the willows along the creek beds as well as all the grass in the park were not spared.

As a result of this, there were no more young aspen trees in the park. The grass was clipped short, and there were no willows along the creek beds. In the absence of willows, the songbirds had no place to build their nests. The beavers that once ate the bark of the willows and aspens and used the wood to build their houses and dams moved out, because they were starving. In the absence of beaver dams, the spring rains caused the creeks to run fast and furious, and the banks of the creeks started caving in. The trout could not stand muddy water, so they also fled.

In desperation, the wardens considered having a harvest of elk. The trouble was that this defeated the purpose of the park.

It was only with considerable difficulty that wolves were brought back into the park. The officials called it "reintroduction." This action is very impressive and also very difficult. Killing is rather easy, but bringing a species back into an area is much more challenging.

The local ranchers were not at all keen on the idea of bringing wolves back into the park, knowing full well that the wolves were almost certain to stray out of the park and attack their livestock. Wolves did not distinguish between elk and cattle.

Despite these difficulties, the wolves were successfully reintroduced.

However, the result was immediate and dramatic. The wolves tore into the herd of elk. The elk that were somewhat feeble of mind and body were the first to fall prey to the wolves. The survivors ran for the hills and began to act like the wild animal that they were.

As a result of this, the grass grew higher. The juvenile aspen trees began to grow again, and the willows grew back. The songbirds returned to build their nests. The beaver returned to the streams, and the water was once again clear, because the beaver dams slowed down the flow of spring water. The banks of the creeks now had willow and tended to stay in place. Of course, the trout returned, because the water was once again clear.

However, the population of coyote took a nosedive.

When people consider the lessons to be learned from Yellowstone National Park, they frequently overlook that little detail. Of course, the coyotes killed more than just young elk, and they were doing their part to destroy the park. The population of the smaller animals like mice, voles, squirrels, chipmunks, and other animals was also being dramatically reduced, and this, too, was having an effect on the park. This effect was just not as noticeable as the others.

Of course the elk were not conscious of the fact that they were destroying their habitat. They just did what they had to do in order to survive.

The wolves were killing the coyotes at every opportunity, not so much for food, but very likely because they saw the coyotes as competition. The wolves kept not only the elk population but also the coyote population in check.

With that in mind, I began to consider the role of the dire wolf.

We now have a surplus of wolves, and the moose population is suffering as a consequence. At least, that is the opinion of the locals where I live, and in fact, none of us are great fans of wolves. Even a small pack of wolves requires a great many moose or elks to survive.

Putting aside my natural prejudice, I acknowledge the need for such predators. I just wonder if the dire wolf kills the timber wolves in much the same way the timber wolf kills the coyotes. This would serve as another example of predators keeping the population of predators under control.

The dire wolf was rarely seen over the years. Now sightings are quite common. Numerous people have asked me about this, and I have no idea why this is the case now. I just know that many forest workers are now spotting this huge wolf on the road between the village and Mackenzie.

Rest assured that forest workers are accustomed to seeing wolves, so most can tell the difference between a timber wolf and something else. That something else is just a dire wolf.

Assuming that the dire wolf kills timber wolves just as the timber wolf kills coyotes and that this serves to keep the timber wolf population under control, just what keeps the dire wolf population under control? I have no idea.

Chapter 11: Bears or Giants

In the spring of 1985, I had my own encounter with bigfoot, a race that I refer to as giants. At the time, I was living alone in a log cabin in the village of Ingenika Point, and my nearest neighbor was a fifteen-minute walk away. Of course, we had no electricity, no roads, and no running water.

At the time, I thought I was going crazy. I could not shake the feeling that I was being watched. That made no sense, because I knew that no one was watching me. The fact was that there were no secrets in such a small village. I just figured that this insanity was a natural result of living alone. It was not. I was being watched.

I should add that a domestic cat had previously dug a small hole under the cabin and crawled in there, a safe place to give birth to her kittens. I also had a fair-sized dog living with me, not a terribly good-natured animal. In fact, he was the meanest dog I could find.

Up north here, we have long days and short nights in the spring.

One day, the sun went down, but it was not quite dark yet. I heard a noise outside the cabin. I was puzzled, because there was only one trail leading to the cabin, and it was easily visible from the window. I had not seen anyone approaching, and the dog was not barking, either.

I went outside, and sure enough, something was making quite a racket; however, it was just out of sight. It was the time of night just before complete darkness. I could see about five yards out, and this animal was at about six yards away. I could not get a glimpse of it. I did get a good look at that dog, and I was not at all impressed. It was terrified! In fact, it was so scared that it was trying to crawl into that small hole the cat had made to find some safety under the cabin.

I thought that dog was a coward, but I was wrong again. That was just a typical response that dogs had to the presence of giants. Dogs do not bark, growl, whine, or run away when they sense these creatures, which is rather strange.

It can also be rather rough on dogs, because I suspect that giants see them as easy meals.

Anyway, I just thought it was a bear, because it was making so much noise and scaring the dog so badly. I called the dog to my side, and for those of you who say that dogs do not make facial expressions, I can only say that you have not seen a dog in the presence of a giant. That was one terrified animal.

Of course, he had better eyesight than I did. If I had seen what the dog had seen, I probably would have joined him trying to crawl into that little hole, too.

I was simply puzzled that this bear was acting to strangely. It sounded like it was walking through clumps of willows and small trees, and bears only did that when they had to. Because I had the place cleaned up nicely just like a park, there was no need for any animal to walk through the willows.

The area around the cabin was nice and open, so why was the bear walking through the willows? Then again, why was it here at all? There was no garbage, no carcasses. There was no reason for bears to come around. It just made no sense.

I decided not to worry about it. Instead, I would just wait until morning and shoot that strange bear in the daylight. There was no way that I was going to grab a rifle and go after any bear this close to dark. This was my idea of being smart, so I went to bed.

Shortly after that, it became completely dark outside, and I could hear that bear on the porch. I was determined to shoot that bear in the morning. I was not worried, because my rifles were loaded and ready, hanging on the wall just over the bed.

Besides, bears always entered a cabin through a window that was closest to the ground. That was the window on the side of the cabin, not the small window higher up at the front of the house. Either way, I went to sleep.

In the morning, I got up, grabbed my rifle, and went hunting for that bear.

I first checked the porch, because bears always left a sign—scratch marks from the claws and droppings and such. The trouble was, however, there were no signs on the porch.

I was completely puzzled, and there was hard-packed ground all around the porch. I checked the ground that the supposed bear had walked across and found nothing. Then I checked the willows that I thought had been walked through and found no broken branches. I was forced to face the only logical conclusion, which was that it was not a bear. It had clearly been a very big animal and so fearsome that it had terrified the dog. Worse, it continued to come around each night after dark, waking me up with the noise it made on the porch. I thought that maybe if I ignored it, it would perhaps lose interest and go away. Such was not the case.

Finally, I went to see the elders for advice.

After they listened to my story, they pointed out that dogs were terrified of giants, which was the reason they did not bark whenever they encountered one. Even though the giants frequently came into the village after dark, most people did not notice them, because dogs gave no warning. The best indication that a giant was around was the dog hiding. They also mentioned that the shaking of trees was typical of giants. Then somewhat to my surprise, I was asked me if I had smelled them. They asked, "Did he stink you out?" Somewhat to their surprise, I mentioned that I had not noticed any strong smell at all.

They then explained how the giants released a blast of foul-smelling air from their rectal orifice in response to an insult. Of course, the elders did not phrase it quite that politely, but it is important. In the interest of politeness, I will refer to this act as "tooting."

If nothing else, this is an indication of their intelligence, their humanity.

Among our species, among all our different cultures and languages around the world, there is one insult that is universal. There is no greater insult than tooting in someone's face. It is the supreme insult, and the giants use it on us. In fact, they use that insult on a regular basis. As for those who object and think that spitting on someone or slapping someone in the face is a greater insult, I can only reply that such acts are assaults more so than insults.

I was then informed very politely that I was a dummy, and I was also in serious trouble. One of the girls who overheard the conversation was not quite so polite. As she put it, because I was living alone, I was "on the shelf." That is a term the girls use for single people, and apparently, even giant girls go hunting for husbands.

As much as I admired that giant's taste in men, I chose to pass on her. Mind you, it does give a whole new meaning to the term "mixed marriage."

It was finally pointed out to me that the stories I had heard—we call them whoppers—were likely based on fact. On occasion, Sasquatch, as they were locally known, did pick up and carry away people. I recalled the other mountain man stories of being picked up and carried away by husband-hunting Sasquatch.

I always thought those stories were just stories, not to be taken too seriously, just a form of entertainment, but after about two months of company on my porch every night, I was not at all sure that it was just entertainment. I was certainly not the slightest bit amused, and I knew that I was being watched all day long. I could think of no other reason that a giant would be sticking so close to me, so I did the smart thing. I listened to the elders and moved away from that place.

Numerous people have also asked me why the giants shake trees. I suspect it is a form of greeting or a warning. The alternative is to simply walk up to our campfires and introduce themselves, but they usually know the response they will receive. They terrify us. We are afraid of the unknown, and we are all too likely to go after them with guns. They are afraid of us, too.

On the other hand, I suspect that their behavior is an indication they want to meet us.

The fact is that the truth is stranger than fiction.

A mountain man by the name of Colter was a famous storyteller. That boy could tell a whopper. Around here, we tend to call them by a more accurate term, even if it is less polite. We call them lies. There is no harm in this. It is entertainment, and it seems to come natural from years of living alone.

Colter was a typical mountain man by all accounts, one not too keen on social graces. He preferred the mountains and solitude to the city life, which I could understand.

Then one day, after he stayed in the mountains for a lengthy time, Colter outdid himself. He came out with a whopper that was an absolute masterpiece. People loved it. It was detailed. It was also graphic, and he swore it was true. The more he swore it was true, the more it was loved.

It never occurred to anyone that his greatest whopper was just the absolute truth, and this was one story that people could not hear often enough.

Colter swore that there was a place where water never froze. The way he told it, even in the middle of winter, the water remained so warm that you could take a bath in it, if you were the sort of soul who indulged in such foolishness. Most mountain men were not terribly concerned with personal hygiene. Colter was no exception.

He also mentioned that there was a place where water came squirting out of the ground as regular as clockwork, day and night, summer and winter. You could set your timepiece by it, if you were the sort of pilgrim to pack such a useless contraption. Mountain men had no use for such devices. They did not need a clock to tell them it was time to eat. They ate when they were hungry just as they slept when they were tired. Any fool knew that it got dark when the sun went down, which made it a good time to sleep. Of course, the sun came up again in the morning, which made it a good time to get up.

Of course, Colter was the first mountain man to set foot on Yellowstone. To this day, Yellowstone is sometimes referred to as Colter's Hell.

His description was detailed and accurate and completely unbelievable. It was also absolutely true. The difference between whoppers and truth is in the details. Whoppers are long on imagination, short on details. The stories I have heard are usually long on details and frequently told by people who are rather short on imagination. I believe these people.

Around 1972, long before there were any roads in this country, a thirteen-year-old girl disappeared from Kwadacha—or Fort Ware, as it was then called.

The RCMP were called in. A search party was organized, and for days, people looked for that girl. After several days, the search was called off. She was given up for dead. No child could survive for very long in the mountains.

After ten days, she walked into the camp of her relatives. She was very much alive and quite healthy. So what had happened?

She explained that a big hairy giant had picked her up and carried her over several mountain ranges to a place where there was a big lake and a run-down cabin. She had not set foot in the cabin, and the giant had fed her raw meat.

From her description, the elders recognized it as a place they called Fox Lake. They were not terribly surprised, because giants could and would borrow people on occasion, especially children. In this case, they figured the giant was likely a female who had lost her baby and was mourning. Borrowing that child was her way of getting a little consolation. Strangely enough, the giants have never been known to hurt children. I consider this one more indication of their humanity.

The disappearance of the girl from Kwadacha was not an isolated incident. A similar event happened at Tucha Lake several years ago when a five-year-old girl went missing.

Of course, the boys went looking for her. After they had searched for several hours, my friend Adam found the girl. She was on top of a hill, and he could not see her through the trees; however, he could hear her clearly telling someone to leave her alone and go away. Adam had a .30-30 lever action rifle, and he levered a cartridge into the chamber. He went up that hill and found that girl alone. When he asked her to whom she had been speaking, she responded, "Spiderman."

It bears repeating that the elders know them well and refer to them as "ice people" or "stink people." Other names include "hairy people" or "giant people." It is significant that all of the First Nations elders to whom I have spoken refer to them as people. Not a single First Nations elder has ever referred to them as animals, at least not to me. In my opinion, these giants are human, merely a different species of human.

The elders have a great deal to say about the giants. First and foremost, they recommend that you leave them alone. This is an excellent piece of advice. They respect us and generally leave us alone, and we should show respect for them and not invade their privacy.

The scientific community is somewhat divided on the subject of the existence of this species. Most scientists are of the opinion that it is absolute nonsense, not to be taken seriously. Anthropologist Jane Goodall is more open-minded on the subject, as expressed in an interview on National Public Radio's *Science Friday* in 2002. She suspects they exist, while her main reservation lay in the fact that we have yet to produce a body. But there is a reason for that.

Several years ago, some people in a different village found a sick and dying giant and cared for that giant as best as they could. This gives an indication of the respect the First Nations people have for these giants. They cared for that giant because of their concern. The giant died despite their finest efforts, and in the middle of the night, the body disappeared.

They think another giant picked up the body in the middle of the night and carried it away. I suspect they bury their dead in caves. This would certainly explain the reason for the lack of physical evidence.

Dr. Grover Krantz, a physical anthropologist, states that *Gigantopithecus* is the largest primate that has ever lived, being considerably bigger than a gorilla and about the right size to fit the description of our North American Sasquatch. This also falls within the range of size that the elders give the giants. They say that most of them are eight feet tall or 2.3 meters while others are over nine feet tall or 2.7 meters.

I suspect this species did not go extinct but evolved bipedal locomotion.

The fact is that all species of human that have ever walked the earth have evolved from apes, and furthermore, that bipedal locomotion is necessary for that transition. These species include the Neanderthal, *Homo erectus*, *Homo habilus*, *Austrolopithecus*, and others. We are not the only species of human to walk the earth. We are just the most successful.

Engels makes it clear that an erect posture was the decisive step in the transition from ape to man. The hand had become free and could henceforth attain ever greater dexterity. The greater flexibility thus acquired was inherited and increased from generation to generation. Engels stresses that the hand is not only the organ of labor but also the product of labor.

But the hand did not exist alone. It was only one member of a highly complex organism. That which benefitted the hand also benefitted the whole body it served. In the words of Charles Darwin, the body benefitted from the law of correlation of growth. This law states that the specialized forms of separate parts of an organic being are always bound up with certain forms of other parts that apparently have no connection with them. Changes in certain forms involve changes in the form of other parts of the body, although the connection may not be readily available. The gradually increasing perfection of the human hand and the commensurate adaptation of the feet for erect gait have, by virtue of such correlation, reacted on other parts of the organism.

Once we establish contact with these giants, a close but discreet examination of their hands will tell us a great deal about their humanity. The closer their hands resemble our hands, the more intelligent, and very likely more human, they are.

I suspect that at some point in the last three hundred thousand years, *Gigantopithecus* evolved bipedal locomotion and made the transition from ape to human. While they evolved in Asia, they probably walked across the Bering Strait when it froze over in the winter.

The elders assure me that the giants moved south for the winter, which helps to explain the absence of tracks in the winter. I suspect they are in the first stage of savagery, members of a hunting-and-gathering society, nomads following the herds. In the wilderness, they are considered one of the top predators in North America, because all other predators, including grizzlies, run from them.

Predators that do not run from the giants can expect a dramatically shorter lifespan. As big as they are, they manage to move like ghosts through the forest. They are rarely seen in the daytime, except in very remote locations.

My friend Allen took a shot at one recently. He was walking in an isolated area, and he thought he saw a bear. That was a mistake. That giant was kneeling down and drinking water, and when my buddy fired, the giant jumped up, looked him in the eye, and ran off into the forest on two legs. Once out of sight, the giant did what they frequently do, even though it is rarely mentioned. He shook a tree.

My friend was terrified, and he ran to the nearest camp, where his buddies were staying. His friends offered to take rifles and go back to have a look. The offer was rather wisely refused, and to this day, he hopes and prays that the giant does not hold a grudge. After all, it was an honest mistake, and we all hope the giant understands that.

Any misguided soul who kills or injures a giant will almost certainly not get out of that forest alive. The giants take care of each other, and if anyone kills or injures a giant, then their friends take action.

All the people who have seen the giants up close agree that the faces are black and shiny. The girls also tend to use such adjectives as beautiful and gorgeous. The fact is that the girls find them very attractive. It is characteristic of males that we tend to size each other up when we first meet, and the fact of the matter is that members of our species are no match for the giants.

Furthermore, of all the reliable sightings of which I am aware, people mostly report seeing male giants. I suspect that the females of that species tend to be more practical and spend most daylight hours in caves, protecting the children.

The experience of Dorothy, who was not a First Nations girl, is rather typical. As a teenage girl at a resort in the southern part of the province, she was told that under no circumstances could she leave the compound and climb the mountain next to it.

The mountain had recently been logged, and the road up the mountain had switchbacks, as we called them. That just meant that the road zigzagged up the mountain. The usual vegetation grew back immediately after it was logged. The vegetation was over waist high by this time.

Is there anything sweeter than forbidden fruit?

In no time, Dorothy and her friend were hiking up the logging road to the top of the mountain. Sure enough, they found trees and rocks that were very similar to the trees and rocks at the bottom of the mountain. They were not impressed. They could not understand what was so special about the mountain. What was so important that they were not allowed up there?

Of course, the girls had no way of knowing that their every move had been watched from the time they had left the resort; however, they soon found out.

As they were walking down the mountain, they approached a switchback. Without realizing the route they were taking, the girls like everyone else walked in the middle of the road on the straight stretches and hugged the corners when they went around the curves.

The giant was just off the road, crouched down on all fours, waiting for them. Just when the girls came to the curve in the road, he jumped up right in front of them. He was so close that he could have reached out and touched them.

The girls took a shortcut down that mountain. They went straight down through the brush and willow, screaming at the top of their lungs.

The giant who had surprised them was likely an adolescent, a teenager like themselves. The girls in the village tell me that they have friends who have experienced similar things. The giants do have a sense of humor, which not everyone appreciates.

Of course, the people at the resort were not the slightest bit surprised. In fact, what had happened was pretty much what they had expected to happen. Dorothy is one of the girls who used the adjective beautiful to describe that giant, although I doubt they had any more dates.

The contrast between the description provided by the girls and that provided by the boys is striking. Both are scared and intimidated by the sheer size of the giants, while the girls appear to find them very attractive. Our mental image of them as dirty, hairy, smelly, rather ferocious, King Kong look-alikes is simply not correct. True, they are hairy, but no girl who has met them up close has mentioned any smell. The girls consider them some of the finest looking men they have ever met. They assure me that if there was a beauty contest for men and if the giants were allowed to compete, then the giants would absolutely win.

It is my hope that Dorothy or some other girl who has a close encounter with one of the giants will sit down with a sketch artist and describe him one day. Because we are unable to get a picture, this is the next best thing.

One of the most intriguing stories the boys have told me is about the time they found some bones several years ago that they could not identify. The bones were very thick, and the boys were puzzled, so they brought one of the bones back to the village and sent it off to a university. I suspect that the bone was sent to the University of Northern British Columbia, but they are not sure where the bone ended up.

From their description, it may just be part of a giant's skull. The only thing they know for sure is that the university called them up and asked if they had any monkeys up here. They assured the university that such was not the case. They have yet to hear any further word from that university.

Without a doubt, we have to make contact with these people. One way to do that is to do the same thing that I inadvertently did a few years ago. We could set up camp in a private area and wait for them to come to us. Their usual greeting is the shaking of trees. We can respond by putting out food for them and letting them know that we are friendly. Obviously, it will take a while to gain their trust.

The problem with that idea is that it could backfire badly. I have been advised that some of those giants are cannibals.

Strictly speaking, we are a separate species, so if members of that species were to eat members of our species, which likely does happen on occasion, then it would not be viewed as an act of cannibalism. Still, it is best to heed the warnings of the elders and scratch that plan. There is a better way to make contact with the giants … or at least a safer way.

My friends from the Reserves located on the coast assure me that the giants love seafood, and because these people respect the giants, they stay off the beach after sundown.

On the mainland, the giants pretty much have to travel south for the winter, not just because they have to follow the herds. Winters here can be quite severe, cold, and deep with snow. That is not a problem on Vancouver Island, because it merely rains in winter for the most part. This is just as well, because the giants are likely stuck on the island, which makes it a logical place to make contact with them.

I am suggesting that we work in cooperation with the First Nations people who live on the coast. The elders could let us know when the giants are in the neighborhood, and we could put some food out for them, perhaps some fruit, vegetables, or meat.

I suspect that after a few meals, the giants will start to get the idea that we are friendly and act accordingly. As it stands now, the giants are mainly active at night, except in remote areas, while during the daylight hours, they can be found in caves; however, it is not recommended that anyone look for giants in caves, because you may actually find them and the women will certainly protect their children by fighting you.

I suspect the giants may likely be found around California or Arizona in the winter. Consequently, this may require a little international cooperation, which could well be a problem. Hopefully, we can keep this a matter of the citizens of two countries working together without all the politics.

I would also like to see laws passed to protect these people. They should feel free to be seen in daylight and live their lives as they see fit. All other species are protected by law. Can we do any less for other humans?

The fact is that they are different from us, and they scare us. We are intimidated by their size. Their strength is awesome. They are huge, and they are hairy. And they do not wear clothes. It is only natural that we fear that which we do not understand, but we are going to have to get over that fear. Those people deserve better than what they have gotten so far.

I am aware that there are people who make a career of tracking down these people, the giants. I am also aware that although their intentions are fine, the result is that they are making the lives of these giants very miserable. These giants are humans, and they have the right to their privacy just as we have the right to our own privacy. They respect us, and we should respect them. We have no right to harass them and attempt to take their pictures, at least not without their permission. Because they avoid cameras whenever possible, that is a fine indication that they are not giving us their permission.

Chapter 12: More on Giants

In the interest of keeping an open mind, I will point out another scientific theory that suggests the Giants may be closely related to the Neanderthals.

It is entirely possible that the Neanderthals may not have gone extinct.

A group of fifty-six international scientists have been examining the bits of bones recovered from a cave in Croatia, the remains of Neanderthal, and comparing the DNA to that of modern humans. They have concluded that our ancestors almost certainly interbred with Neanderthal.

This is considered scientific proof that the Neanderthal did not so much go extinct as merge with *Homo sapiens* and that we are in fact a mixture of two species, Neanderthal and *Homo sapiens.* With that in mind, it is entirely possible that a separate branch of that species of human has evolved into what we now refer to as Sasquatch.

Be that as it may, the giants are definitely not extinct, and I suspect they may be just as intelligent as we are. They have managed to avoid us quite nicely, and most members of our species are unaware of the existence of these people, even though they are well aware of our existence.

From all accounts, their tools are rather primitive, at least compared to our tools. These tools may not accurately reflect their intelligence as much as their incredible strength. I suspect they have developed the tools they need to survive just as our species has developed the tools we need to survive.

The only difference may be that we need far more sophisticated tools to survive, at least in most parts of the world, because our species is so puny, compared to the giants. And rest assured, the bow and arrow is a very sophisticated tool. I am not even sure the giants use knives, probably because they do not need them. With their great strength, they are quite capable of butchering an animal with their bare hands.

It is just a matter of time—and very likely a very short time—before we have a serious run in with the giants. Certainly, their behavior is changing.

Recently, in Kwadacha, some young people were partying by the airport. They were in pickups, and it was getting dark. Then they noticed a giant coming closer and closer to them, trying to hide behind some clumps of willows. They fired up those vehicles and got out of there in a rush. Apparently, the giant could not resist the lure of alcohol.

Although most people do not bother, there is nothing easier than making alcohol. The only thing required is sugar, water, and some kind of fruit or vegetable for flavoring, and the sugar in the fruit or vegetable adds to the alcohol content. It also provides a certain flavor, which at times can be a mixed blessing. By that I mean that the taste of that brew can be rather vile.

The point is that a little yeast gets the process off to a roaring start. Then it is just a matter of keeping the mixture warm for a few days, and the resulting fluid can be quite potent.

Around here, we call it "home brew" or just plain brew. It is just one more name for alcohol, and though it is not as potent as moonshine, which is made by means of a different process, it can still get one very drunk.

I personally know a couple of people who each drank about two liters of a batch and each managed to get so drunk that both blacked out and couldn't remember what had happened the night before. It was clearly the best batch of brew they had ever made.

Until very recently, some people here used to set home brew in isolated cabins and leave it for a few days but not anymore. By the time people came back, the brew was gone.

At first they were puzzled, thinking perhaps that someone had stolen their brew, but then they noticed the hair that the giants had inadvertently left behind.

Apparently, the giants know about alcohol, and they love the stuff. They are truly human. They love the same things we love. Sadly, that includes alcohol, not that I blame them in the slightest.

During hunting season, hunters tend to go into those forests with something more than their rifles. Of course, I am referring to alcohol. It is just a matter of time before one of those giants is going to get his hands on a bottle and get drunk.

God help anyone who gets close to a drunken giant.

Recently, three young people in the village of Tsay Keh Dene saw a giant. It happened during a summer evening, and the three were on quads. The giant was thirsty and went to the lake to get some water. It is a man-made lake, and the water is used to generate electricity, so the level of the water fluctuates rather wildly.

During this particular time, the lake was quite low, with a rather steep drop. The giant faced the problem and quite simply pushed over a fair-sized tree and used that to slide down to the lake. This gives some indication of the strength of those people.

The giant got down on one knee and started scooping up water with one hand. The young people sat there for five or ten minutes by their estimate and watched him. He must have known they were there, especially because they sounded the horn for him. The giant was probably older, judging by the gray head and gray torso. After he was through drinking, he stood up and walked down the lake shore.

To be sure, the boys from the village went to that particular spot and looked for tracks the next day. They found nothing, which did not surprise me. The giants obviously know how to hide their tracks.

The fact that the giant was not at all concerned that the kids were watching him is an indication of the change of attitude of those people. I suspect they are resigned to a confrontation with our species.

During another incident years ago, some of the boys shot a moose at Akie Swamp. It was a preferred hunting ground, a fine place to shoot moose. It was the local equivalent of a supermarket for local hunters.

As happens on a regular basis, a moose was shot there one day close to dark. The boys just had time to butcher the moose as it was getting dark, and they decided to come back in the morning. This was something they had done on numerous occasions in the past. And just as they had done countless times before, they returned in the morning with some dogs to pack the meat.

On that particular day, things did not work out quite as planned. In the morning, every little bit of the moose was gone. Nothing was left. Even the hide was missing.

What likely happened is that a giant had happened along and decided to help himself to the spoils. Clearly, not all giants are the bashful sorts. This one saw an opportunity and seized it.

I suspect we have meddled in their food caches, so turnabout is fair play. One good turn deserves another. I do not blame him in the slightest, but then, it was not my moose. It is easy to be generous with what does not belong to us.

The fact that the giant was able to pick up a moose, put all the meat into the hide, and pack it away is another indication of the strength of these people.

There are other indications of their humanity, too.

Recently, Elsie and the girls in the village were picking berries when they found something strange. Someone had placed a great many branches of huckleberries there in the clearing. They were piled in neat rows, not thrown down randomly. Of course, the berries were ripe, and the branches were covered with them. It is significant that only the branches had been removed. Clearly, no berry plants had been killed. They were piled very neatly in stacks about knee high. The girls knew better than to touch them. The clearing was facing south and was sure to get a great deal of sunlight. This is probably one of the ways that giants preserve their food.

It is my impression that the giants are respected and admired by the people to whom I have spoken. Most consider it good luck to see a giant. As long as giants are close to you, you certainly need not worry about predators. When people are out camping and giants are around, they usually put the rifles away and put out food for these creatures. It makes sense to be nice to giants and feed them. As long as the giants are around, we need not fear other predators.

Chapter 13: Industrial Developments and Border Fences

Here in the village, there is talk of setting up a dam on the Finlay River to provide this village as well as other places with electricity. This would have the advantage of freeing the village from its dependence on generators. This idea, as well as others, are being discussed. As yet, it remains merely an idea.

Of course, the demand for electricity is never-ending, and I certainly enjoy flicking a switch as opposed to lighting a lamp. I suspect the area that could be flooded includes Buffalo Head Mountain, a perpendicular mountain, which is one of the nesting grounds of the pterodactyl. If so, it could well throw a monkey wrench into the works.

The other item that caught my attention was the plans for border fences along the Canadian-American borders as well as the Mexican-American borders.

The woolly mammoth is a very large animal, and we should be able to spot it from satellite cameras or planes. The only reason it has not been seen until now is because no one has looked for this creature. As soon as they come out of the caves in the spring and someone looks for them, they should be spotted. After they are proven to exist, they will no doubt be placed on the endangered species list.

The mammoth will sense the pressure is off them and will likely return to the prairie. As an endangered species, they will come and go as they please. It is very likely that their current practice of spending the winter in caves was forced upon them after they were chased out of the prairie. I suspect that in years past, the mammoth travelled south for the winter in order to avoid the severe cold and deep snow. Do not be surprised if they revert to their old habits and once again travel south for the winter. In that process, they will be eating and trampling the crops of farmers as they move. Most of us prefer California sunshine to Canadian winters, so this is understandable.

I can just imagine the fun those animals will have on the golf courses.

This will also mean that the sale of mammoth ivory will be banned. Such sales are currently legal, but only because the mammoth is considered an extinct species. As for those who think it is possible to hunt the woolly mammoth before it is placed on the endangered species list, such is not the case. It is currently illegal to hunt any animal not mentioned on the hunting regulations, and rest assured that the mammoth is absolutely not mentioned. That at least is true in this province, and I am sure it is the law elsewhere as well. At least I hope it is.

So too, as soon as we prove that giants exist and that they are a separate species of human who are citizens of no country, they will be protected by the United Nations and will be allowed to come and go as they please, able to pass through international borders.

Without a doubt, they pass through the Canadian-American border without bothering with any checkpoints. It's also likely that they pass through the Mexican-American border in the same way. This is not about to change.

The only difference is that they will soon be protected by law.

We can forget about fences at the international borders, at least in the areas where the giants or the mammoths travel. The fact is that any and all man-made obstacles to the migration route of the giants or the mammoths can and will be removed. No further obstacles will be placed in their way.

It is quite likely that the mega bear, the dire wolf, and the plesiosaur will also be placed on the endangered species list. Assuming that is the case, then their habitat will have to be protected. This will have an effect on industrial development, such as logging and mining. This includes the lakes in which the plesiosaur lives. It is quite possible that the current practice of placing hundreds of truckloads of logs in the water and booming them to the mills may have to change. A great deal of bark falls off the logs and causes pollution, which may poison the animals.

This is not to say that industrial development should end, just that it will have to take place with consideration given to these species.

There remains the problem of proving the existence of these species. It should not prove terribly difficult, if we first consult with the experts, the First Nations elders. We have got to combine the wisdom of the First Nations elders with our knowledge of modern science. I am sure that each can learn a great deal from the other. The way to do this is really quite simple and easy. It is all a matter of showing each other the proper respect.

The correct approach is to call the Reserve and request a meeting with the elders. Most Reserves have an elders society. The elders may or may not agree to the meeting, but it has been my experience that when people are treated with respect, they tend to respond with respect. On the other hand, when people are treated with contempt, they tend to respond with contempt. Who can blame them?

Assuming the elders agree to a meeting, then the band can set up a meeting. It is up to the people requesting the meeting to provide a proper feast.

The people in charge of band administration can make the arrangements or tell you what they expect. The food they love will vary from location to location. Some people love salmon. Others may love moose meat. Still, others may enjoy fried chicken. Spare no expense. Provide them with variety and lots of it.

In addition, each elder should be provided with a gift, such as tobacco or a Hudson Bay blanket. If an elder is too feeble to attend the meeting, then have these gifts delivered to their homes.

As long as the traditional territory of the band is in these mountains, then rest assured that these people are the experts on the mountains. After the meal and formalities are over, it is then time to get down to business. You can ask them about the animals mentioned in this book and ask for advice on the best way to find them. I am convinced that this is the way forward.

The young people need to be involved in this science project, too. They will learn not only the proper scientific method but also how to respect the elders. As it stands now, many youngsters scoff at the stories that the elders tell and even the stories of their friends. They quite honestly do not believe these species exist. When they personally have an encounter with these species, it is quite a shock. George can testify to that.

Several years ago, when George was just a teenager, he stood in the school yard of the village just across the road from the village school. He started walking home with his sister, Elizabeth, and his cousin, Deanna. It was dark outside, but there was a full moon that night, too. The sound of the flapping wings caught his attention, and immediately, the stories he had heard while he had sat on his grandmother's knee came to mind.

George yelled a warning to the girls, and they hit the ground. They could see the devil bird coming at them, flying close to the ground, its legs outstretched, talons extended. They could feel the breeze created by the flapping wings as it flew past, and they would hear the snapping sound as the claws slammed shut on empty air, the predator cheated of its meal, the animal barking in frustration.

The teenagers watched in shock and disbelief as the devil bird flew directly for the store and lifted up at the last instant and soared over the building. They all swear that the wings were as wide as the store overhang, 8.6 meters or twenty-eight feet, and that the tail was six feet long. Only then did they realize that the elders were right. These species exist. We just have to find them.